SPRINGER TRACTS IN MODERN PHYSICS

Ergebnisse
der exakten Natur-
wissenschaften

Volume 59

Editor: G. Höhler

Springer-Verlag Berlin Heidelberg GmbH 1971

Manuscripts for publication should be adressed to:

G. Höhler, Institut für Theoretische Kernphysik der Universität, 75 Karlsruhe 1, Postfach 6380

Proofs and all correspondence concerning papers in the process of publication should be addressed to:

E. A. Niekisch, Kernforschungsanlage Jülich, Institut für Technische Physik, 517 Jülich, Postfach 365

ISBN 978-3-662-15879-1 ISBN 978-3-540-36661-4 (eBook)
DOI 10.1007/978-3-540-36661-4

Originally published by Springer-Verlag Berlin Heidelberg New York in 1971.
Softcover reprint of the hardcover 1st edition 1971

Symposium on Meson-, Photo-, and Electro-production at Low and Intermediate Energies

Bonn
September 21—26, 1970

Contents

Pion Photoproduction on Nucleons in the First Resonance Region (Experimental Situation)

G. von Holtey

Contents

I. Introduction

The intention of this talk is to outline a picture of the experimental situation of the photoproduction of single pions on nucleons in the first resonance region.

Pion photoproduction from nucleons has been, for nearly two decades, the subject of a great deal of attraction, both experimental and theoretical. Up to 1966 more than 50 experiments have been reported measuring photoproduction reactions in the resonance region*.

These experiments confirmed the theoretical predictions of a predominant magnetic dipole exitation in π^0 photoproduction and big contributions from both the magnetic dipole and the electric dipole exitation in π^+ photoproduction.

However, in spite of the numerous experimental information, the data were not sufficient to allow for a detailed calculation of the smaller multipoles. Comparing the different measurements, available at that time, one finds a broad band of deviations typically of the order of 20–30%.

In the last few years the theoretical understanding of the photoproduction process has considerably improved by calculations, using dispersion relations as well as the isobar model [2–6]. In most of these theories phenomenological parameters occur, which have to be determined by fits to the experimental data.

* A complete compilation of the data up to 1966 has been given by *R. L. Walker* [1].

In this situation, where accurate and reliable normalised data were highly required, new experiments were performed at several laboratories, using refined equipment. With these new measurements the experimental situation has considerably improved, allowing, for example, for a consistent multipole analysis in low energy photoproduction [7, 43].

In the following, I want to discuss these new experiments with the main point on a comparison of the obtained data. Thereby, only measurements with unpolarized photons will be considered. A review of the recent experiments with polarized photons has been given by *G. Bologna* [42].

II. Photoproduction of π^0 Mesons on Protons

To begin with the π^0 photoproduction on protons, as pointed out by *Walker* [36], this is a reaction with a well established tradition of disagreement between different experiments. However, the situation has considerably improved in the last few years mainly by the experiments at Bonn and Orsay.

One of the main difficulties of photoproduction experiments in the first resonance region are the numerous error sources contributing in the same order of magnitude and, moreover, in most cases show up a

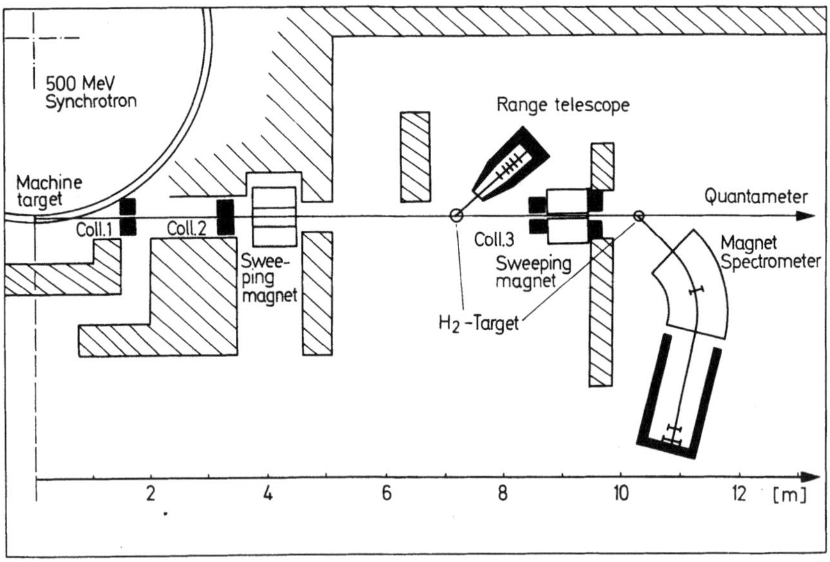

Fig. 1. Experimental set-up of the two simultaneous π^0 photoproduction measurements at the Bonn 500 MeV synchrotron

strong dependence on kinematical variables. In this situation a reliable estimate of the systematic errors becomes rather difficult.

In order to overcome this obstacle and to reduce the total systematic error down to ±5% the Bonn group performed two simultaneous experiments on π^0 photoproduction using two independent experimental techniques for detecting the recoil proton [8]. Fig. 1 shows the two experimental set ups, a range telescope and a magnetic spectrometer, both using the same gamma-ray beam of the 500 MeV synchrotron.

Differential cross sections were investigated for laboratory photon energies between 200 and 440 MeV and for center-of-mass angles larger than 50°, permitting a direct comparison of the two data sets in a wide kinematical range. In this manner the influence of systematic error sources inherent in the different experimental arrangements could be studied very carefully. An example of these data is shown in Fig. 2.

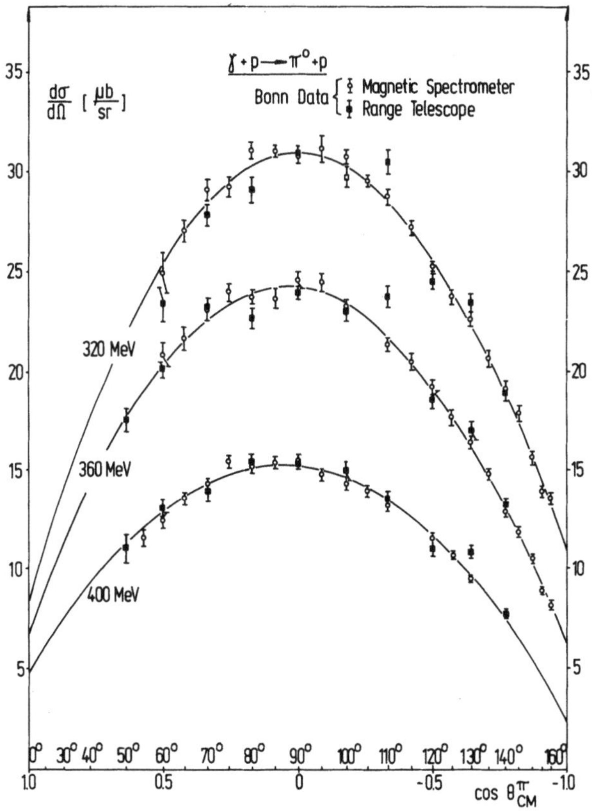

Fig. 2. Angular distributions of the π^0 photoproduction differential cross section for various energies. ⧲ magnetic spectrometer data, ▮ range telescope data, full line: polynomial fit to the spectrometer data

For this comparison polynomial fits to the magnet data were used as references in order to suppress the influence of counting statistics. The resulting systematic deviations between the two experiments are shown in Fig. 3. They vary between 0% and 7%, demonstrating a clear

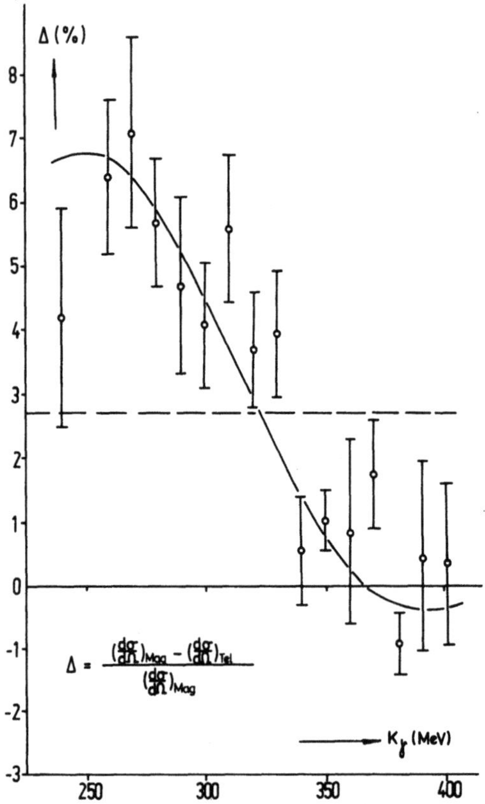

Fig. 3. Deviations between the Bonn magnet and telescope cross sections. The deviations are averaged over the angular distributions

dependence on photon energy. This can be understood by assuming a 0.6% shift in the photon energy calibration and a 2.6% scale factor difference between the two experiments. It should be stressed, however, that these deviations are smaller than the total systematic errors, which have been estimated to be less than ±5% for both experiments.

The two measurements therefore were combined to one data set by correcting the scale factors and energy calibration of both experiments in equal parts. As an example Fig. 4 shows some of the 23 obtained

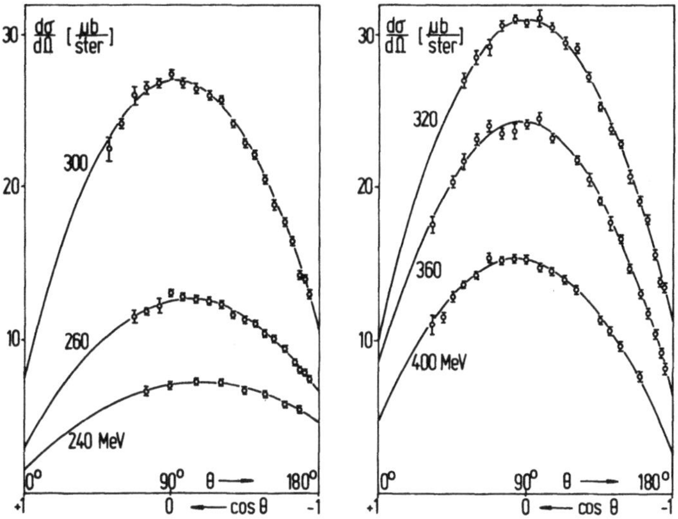

Fig. 4. Angular distributions for various energies of the combined Bonn π^0 photoproduction data. The full line is a polynomial fit to the data

angular distributions of this combined data set. The curves in Fig. 4 are second order polynomial fits in $\cos\theta$.

Below 450 MeV photon energy, the results of theoretical analysis indicate contributions of *s*-, *p*- and *d*-waves to the π^0 differential cross section, thus necessitate polynomials up to the fourth order in $\cos\theta$ for fitting the angular distribution, i.e.

$$\frac{d\sigma}{d\Omega} = A + B\cos\theta + C\cos^2\theta + D\cos^3\theta + E\cos^4\theta\,. \qquad (1)$$

Various theoretical predictions yield *D* and *E* coefficients less than 1 μ b/ster in the considered energy region, that is in the order of the *B* coefficient only. The determination of the coefficient *D* and *E* from the Bonn experiment, however, is not possible, mainly because of the restriction in the kinematical region to center-of-mass angles larger than 60°. By introducing third and fourth order coefficients to the fit the χ^2 is not improved, however, extrapolations outside the experimental region are strongly affected. For instance, extrapolations to zero degree vary by about 25% for fits with or without *D* and *E* coefficients.

In the next figure (Fig. 5) the energy dependence of the *D* coefficient is shown as obtained from theoretical calculations by *Schwela* [4], using dispersion relations, and by *Pfeil* [5] using the isobar model. The experimental values are taken from third order polynomial fits to the Bonn π^0 data [8]. Although both authors essentially used the Bonn data for the determination of the phenomenological parameters the results are quite different as far as the D and E coefficients are concerned. This clearly demonstrates the need of accurate data at forward angles in the region of the first resonance.

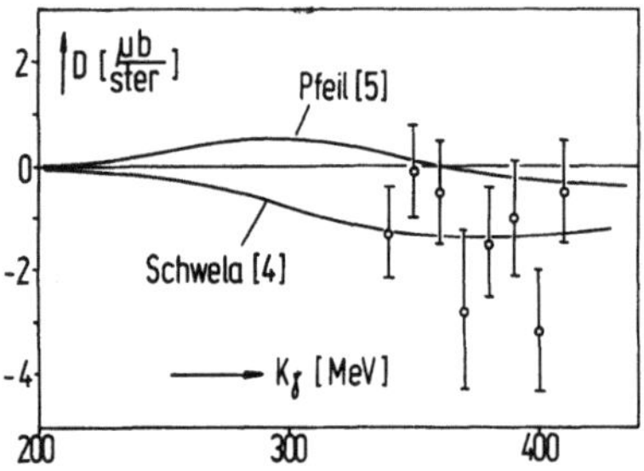

Fig. 5

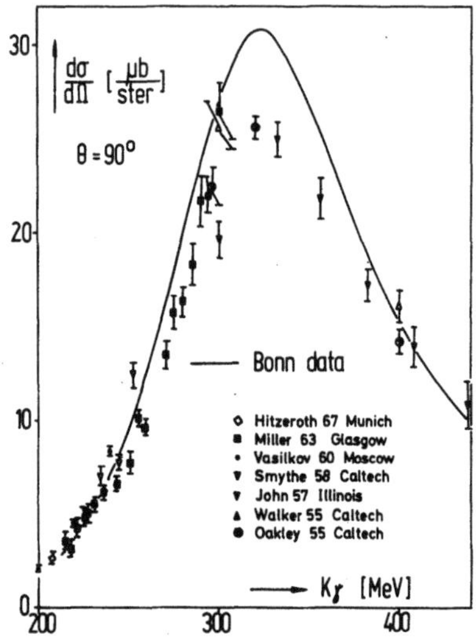

Fig. 6

Fig. 5. The coefficient D of the polynomial fits to the π^0 photoproduction data. The two curves show theoretical calculations (see Ref. [4, 5]). The points are due to fits of the experimental Bonn data

Fig. 6. Comparison of the Bonn data (full line) with other measurements at $\theta_{CM} = 90°$

For a comparison of the Bonn π^0 data with previous measurements the 90 degree excitation curve, where most experiments have been done, is drawn in Fig. 6. On inspection one finds clear evidence for the fact, that the Bonn data are systematically 10–15% higher than nearly all previous measurements. The resolution correction, which is missing in most of the previous experiments, could possibly explain at least part of this discrepancy.

Let us now turn to the Orsay experiment on π^0 photoproduction [9]. The Orsay group measured exitation curves from 220 to 400 MeV at four angles between 70° and 175°. The recoil protons were analysed by a double focussing spectrometer. The acceptance of the detection system was normalized to elastic electron-proton scattering. As an example Fig. 7 shows some of the obtained data.

The agreement between this experiment and the Bonn π^0 data is reasonably good. The deviations, which vary between 0% and -10% with respect to the Bonn data can be understood by a 5–6% scale factor difference, which is compatible with the estimated systematic errors.

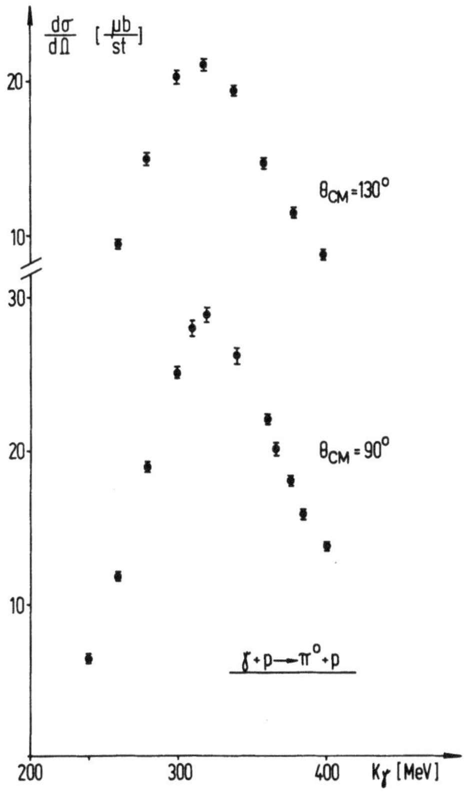

Fig. 7. Exitation curves for two angles of the Orsay data, Ref. [9]

A rather sensitive measure of the agreement between π^0 photoproduction measurements is the fit coefficient B, which gives the foreward-backward asymmetry. In Fig. 8 it is drawn, as obtained by fitting the data of the two experiments separately. The large errors of the Orsay B values are due to the fact, that here the sum of statistical and systematical errors of the data were used for fitting [9], whereas in the Bonn data fit only the statistical errors were used. Again the agreement is quite good. Moreover, this picture clearly shows the zero-crossing of the coefficient B near the resonance.

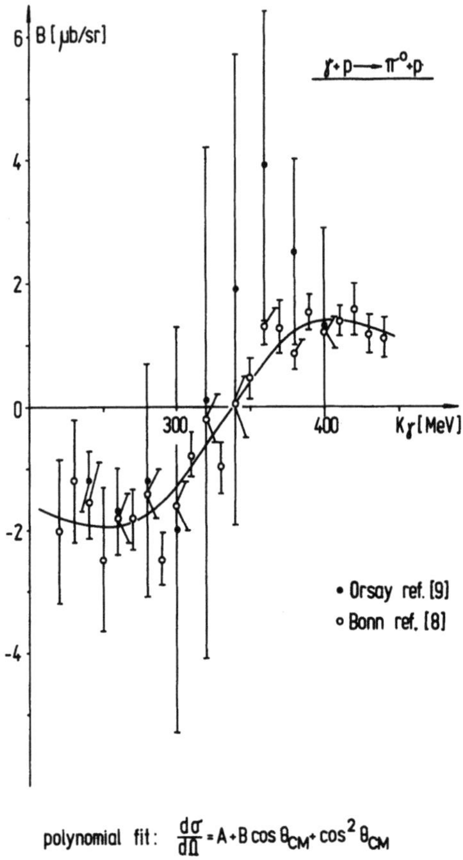

Fig. 8. The coefficient B of the polynomial fits to the π^0 photoproduction data. ⊕ Bonn data, Ref. [8], ● Orsay data, Ref. [9]

With these two new measurements the experimental situation for the photoproduction cross section of neutral pions is quite satisfactory in the energy region around the resonance and for centre-of-mass angles larger than 60°.

In order to extend these measurements to smaller angles an experiment has been performed very recently at Bonn, detecting the two photons from the π^0-decay with total absorption Čerenkov counters [10]. In this experiment angular distributions between 10° and 70° were taken for photon energies between 340 and 420 MeV. The results at two photon energies are shown in the next figure (Fig. 9) together with the Bonn and Orsay data. The solid line represents the fit of the recent multipole analysis by *Noelle et al.* [7].

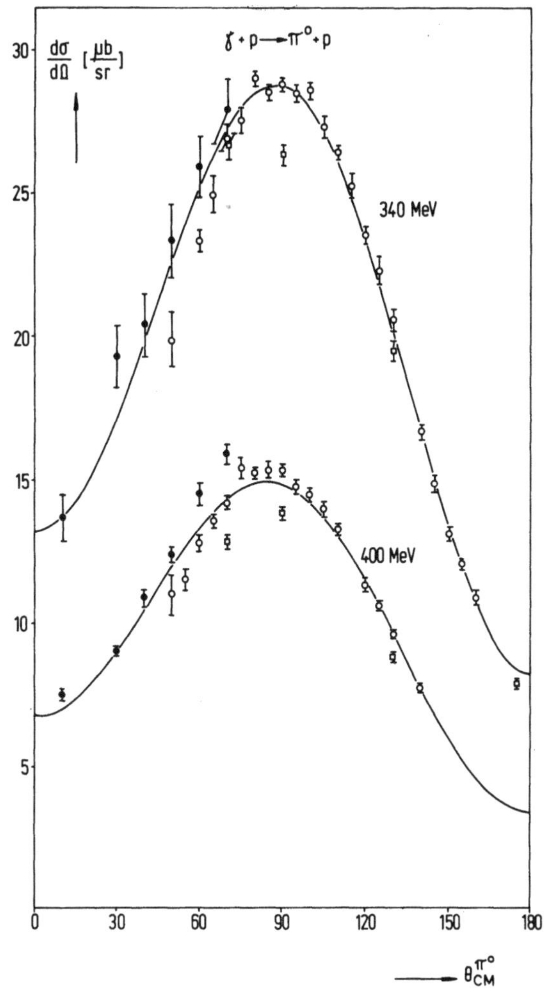

Fig. 9. Comparison of the differential cross section at two energies on π^0 photoproduction of different recent experiments. ● Bonn, Ref. [10], ○ Bonn combined data, Ref. [8], □ Orsay, Ref. [9], full line: multipole analysis by *Noelle et al.*, Ref. [7]

The comparison indicates, that the new experiment gives typically 10% higher values than the combined magnet-telescope measurement. However, in spite of this, the two measurements are compatible with respect to the total systematic errors. In the overlapping region, the largest systematic errors of the combined Bonn data are expected, because of the low momentum of the detected proton.

At present a further experiment is in progress at Bonn, with the aim to measure the π^0 photoproduction cross section at small angles and low energies, detecting the recoil protons coming from a gas target [11].

At energies near threshold there are several measurements [12], as illustrated in Fig. 10. They all agree reasonably well and also give a smooth extension, at least at large angles, to the mentioned Bonn and Orsay results. Only few points exist in the kinematical region $\theta_{CM} < 90°$; 200 MeV $< k_\gamma <$ 280 MeV and in addition, the extension of these data to the threshold measurements is not very convincing.

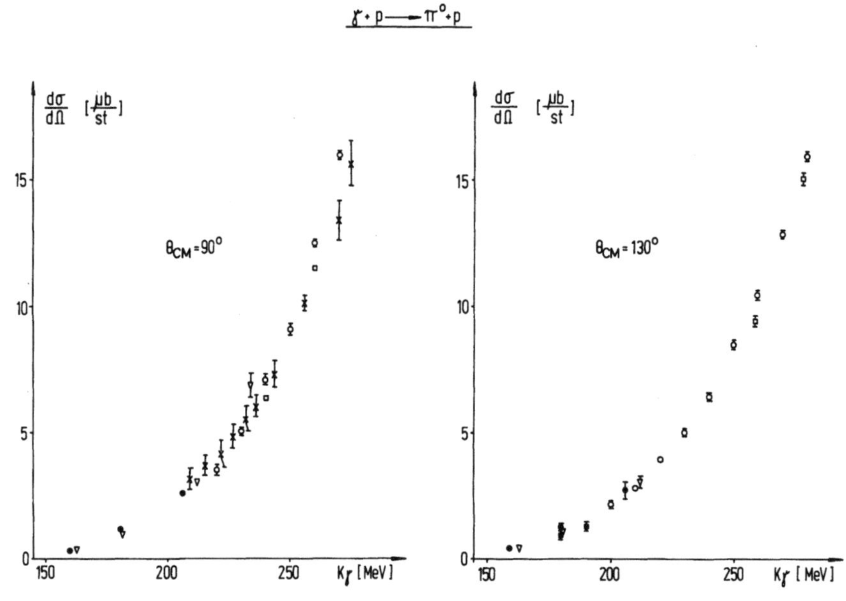

Fig. 10. Energy dependence of the π^0 photoproduction differential cross section near threshold at two angles. The experimental points are labelled: ϕ Bonn, Ref. [8], ⌶ Orsay, Ref. [9], ● *Hitzerroth*, Ref. [12a], ∇ *Govorkov et al.*, Ref. [12b], × *Miller et al.*, Ref. [12c], ■ *Koester et al.*, Ref. [12d]

At energies above 440 MeV π^0 photoproduction data are still scarce and not very accurate. These energies, however, are very important for the determination of the energy dependence of the multipoles between 400 and 500 MeV. In the recent multipole analysis [7], for example, both isospin parts of the M_{1-} multipole turned out to be completely undetermined in this energy region.

Coming now to the measurements of the recoil proton polarisation in neutral pion photoproduction, Fig. 11 shows all available data in

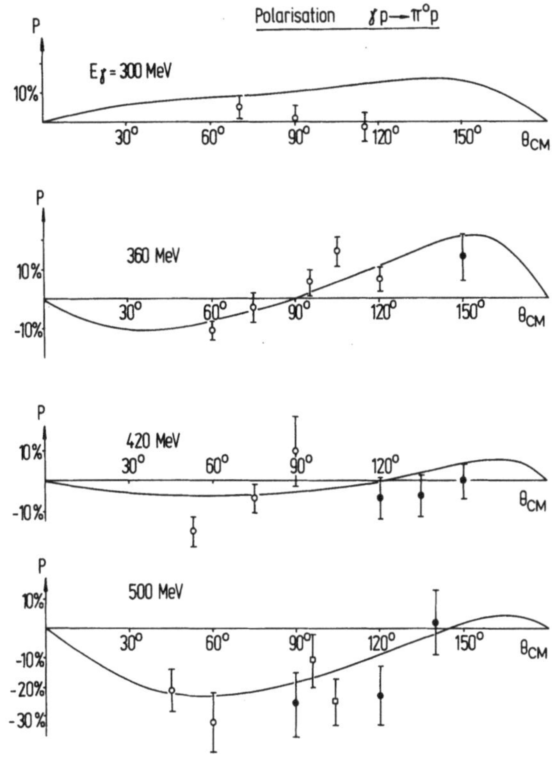

Fig. 11. Polarisation of the recoil proton in the reaction $\gamma p \to \pi^0 p$. Open and full circles are Bonn measurements with He respectively C_{12} analyser (see Ref. [13]). Open squares are points from Stanford, Ref. [14]. Full line: multipole analysis by *Noelle et al.* [17]

the region of the first resonance, except two points at 240 MeV and 250 MeV. Most of the points are yielded at the Bonn synchrotrons, using He and Carbon (C_{12}) as an analyser [13]. At 500 MeV, and higher energies, there are also measurements from Stanford [14] and Tokyo [15] both using a carbon analyser. These data fit rather well into the Bonn polarisation measurements.

III. Photoproduction of π^+ Mesons on Protons

Let me now turn to the photoproduction of positive pions on protons. The situation for the differential cross section without any polarisation is very similar to that described in the π^0 production case. Here again the experimental situation has been considerably improved in the last few years, mainly by experiments at Bonn, Orsay and Moscow.

The two measurements of *Adamovich et al.* [16] from the Lebedev Institute yielded a considerable amount of data in the threshold region, taken with the emulsion technique. The Orsay group [17] measured differential cross sections in a wide kinematical range for c.m. angles between 0° and 130° and for photon energies between 300 MeV and 750 MeV. They again used a double-focusing magnetic spectrometer. The Bonn measurements [18] were performed with the same magnetic spectrometer as was used for the π° experiment described above. Here, more than 400 differential cross sections were obtained at kinematical

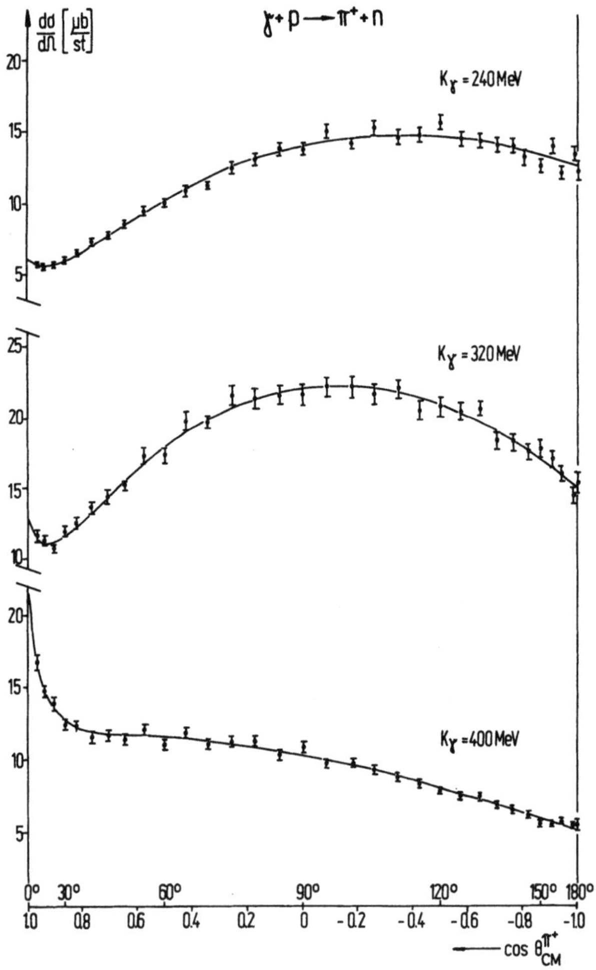

Fig. 12. Angular distributions at various energies of the Bonn π^+ photoproduction differential cross sections, Ref. [18]. The full line is a least square fit to the data

settings between 15° and 180° c.m. angle and between 220 MeV and 425 MeV photon lab. energy. The estimated systematical errors of the Orsay and Bonn data are less than ±5% respectively ±6%.

I shall begin with the Bonn π^+ data for a more detailed description, mainly because the angular fits to these data provide a good basis for comparison purposes.

In Fig. 12 some typical angular distributions are shown obtained by the Bonn group. The solid line is a least square fit to the data. The specific form used in fitting is the following

$$\frac{d\sigma}{d\Omega} = \sigma_B + a\frac{1-x^2}{1-\beta x}\cdot x + \sum_{n=0}^{2} b_n x^n \qquad (2)$$

where $x = \cos\theta_{CM}$ and σ_B is the differential cross section following from the electric part of the Born amplitude. The fits were done with a fixed coupling constant ($f^2 = 0.08$). The second term takes into account for the interference between the Born amplitude and the resonant part. The choice of this formula has the virtue of clearly exhibiting the resonance behaviour by the sum. This is shown in the next figure (Fig. 13), where the π^+ total cross section obtained by integrating only the resonance term of the fit, i.e. the third term of the r.h.s. of Eq. (2), is plotted together with the π^0 total cross section divided by two. The two curves nearly coincide, as they should by assuming a pure 3/2-isospin resonant state. The difference is due to some nonresonant background and to

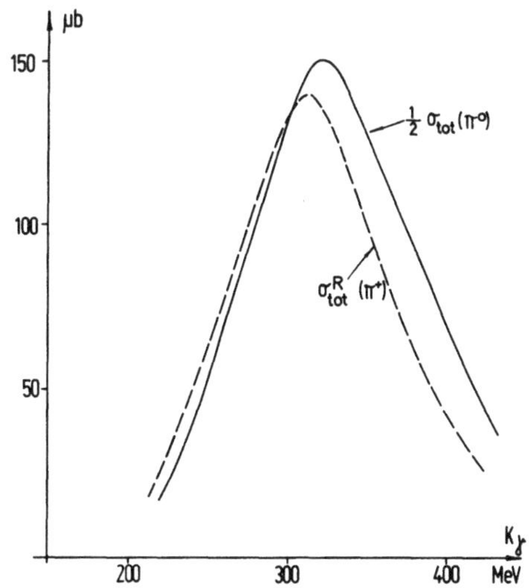

Fig. 13. Comparison of total cross sections. Full line: total cross section of the reaction $\gamma p \to \pi^0 p$ divided by two; dashed line: resonant part of the total cross section of the reaction $\gamma p \to \pi^+ n$ (see Eq. (2))

a small rest of the interference between the separated pole term and the resonant amplitude.

Moreover, these fits offer a reliable representation of the Bonn π^+ data combined with a strong suppression of counting statistics and therefore, as already mentioned, provide a good reference for comparison purposes.

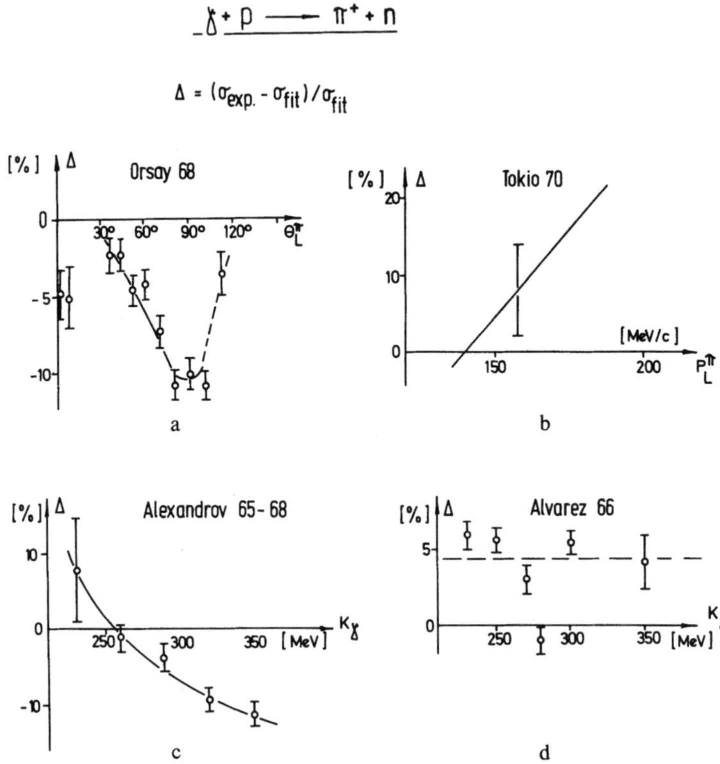

Fig. 14. Deviations of recent π^+ photoproduction experiments relative to the fit values of the Bonn measurement Ref. [18]. a) *Bétourné et al.*, Ref. [17]; b) *Fujii et al.*, Ref. [21]; c) *Y. M. Alexandrov et al.*, Ref. [19]; d) *M. A. Alvarez*, Ref. [20]

This is done in Fig. 14, where the deviations of most of the recent experiments with respect to the Bonn data are plotted. In all these cases no deviations were found, which showed a systematic dependence on other variables as indicated here.

To begin with the Orsay measurement [17] (Fig. 14a), a clear dependence of the deviations on the pion lab. angle exists showing a maximum at 90° (the deviations are averaged over the exitation curves).

I should point out, however, that there is only a small region with deviations larger than 5% and even in this region the measurements are compatible with respect to the stated systematic errors; that is to say, the overall agreement between the two experiments is fairly good.

Next there are the data of *Alexandrov et al.* [19]. These authors obtained angular distributions at five photon lab. energies between 230 and 350 MeV in two different experiments. The deviations averaged over the angular distributions show a pronounced dependence on the photon energy (Fig. 14c). However, as these are results from two different experiments using quite different detection methods, one should not overvalue this picture.

The experiment of *Alvarez* [20] at the Stanford Mark III linear accelerator yielded angular distributions at photon energies between 225 and 350 MeV. These data were normalized to an absolute cross section near the peak of the resonance, measured by means of a polyethylene-carbon substraction using solid targets. The deviations of this normalization point in respect to the Bonn data fit is indicated by the dashed line in Fig. 14d. Consequently, besides this normalization effect, there is a remarkable good agreement between both experiments.

Finally, the averaged deviation of the very recent data at backward angles from the Tokio Institute [21] is indicated (Fig. 14b). These data are preliminary. They show a distinct increase in respect to the Bonn π^+ backward measurements with increasing pion lab. momentum. This characteristic feature becomes even more pronounced with higher photon energies.

What is not shown here, is a comparison with the previous Bonn telescope measurement of the π^+ photoproduction cross section [22]. There are large deviations ranging from plus to minus 20% with a very pronounced dependence on the pion lab. momentum. This disagreement is most probably due to an incorrect absorption cross section, applied to the nuclear absorption correction, which – in the case of the telescope – amounts up to rather high values. An effort in account to correct the data for this effect is in progress [23].

The extension of the hitherto discussed π^+ data to the threshold region is illustrated in Fig. 15. Here the low energy measurements of *Adamovich et al.* [16] are plotted together with the Bonn and Orsay data. The solid line – a hand drawn curve through the polynomial fit values of the Bonn measurement – is smoothly extrapolated to the experimental threshold value, $\left(\frac{k}{q}\frac{d\sigma}{d\Omega}\right)_{th} = 15.6 \pm 0.6\ \mu b/st$, measured by *Burq* [37]. This experimental value is in good agreement with recent theoretical calculations following the chiral lagrangian model [38] as well as current algebra [39]. At 90° the extension is rather smooth,

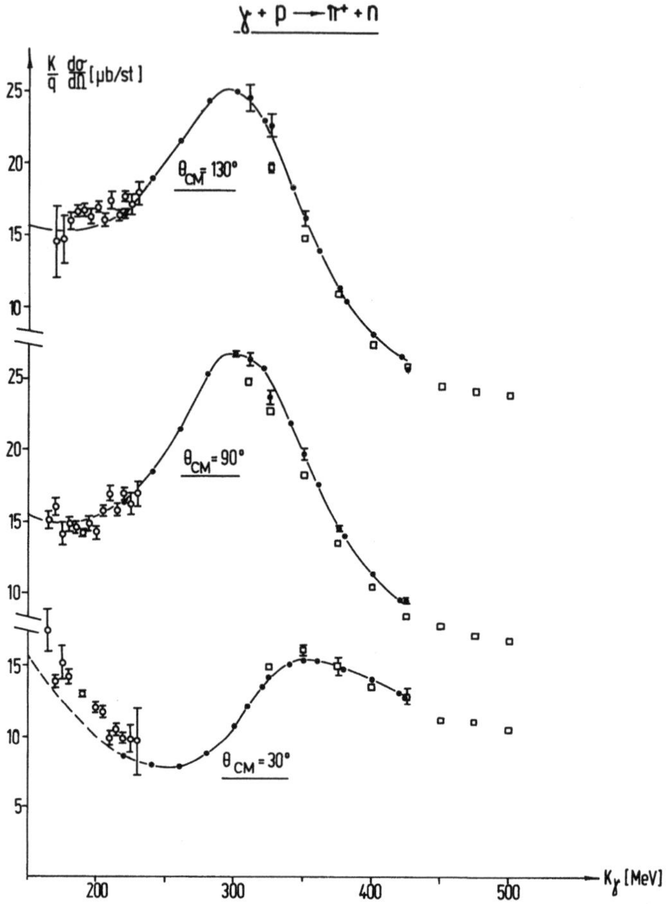

Fig. 15. Comparison of the π^+ photoproduction data near threshold with data in the resonance region. The points are polynomial fit values to the data. ⌽ Moscow, Ref. [16], ⧳ Bonn, Ref. [18], ⌶ Orsay, Ref. [17]. The solid line is a hand drawn curve through the polynomial fit values of the Bonn experiment

the same is more or less true for higher angles. For small angles, however, the transition to the low energy part is not at all smooth. There are systematic deviations in the order of 20%. However, having in mind the difficulties of the emulsion experiment and the limited statistical and systematical accuracy of the obtained data, the overall agreement is still reasonably good.

To summarise, the experimental situation for the π^+ photoproduction cross section in the resonance region is quite satisfactory now. There are accurate measurements in the whole energy region, here under consideration, which cover the total angular region. This, once more, can be seen in Fig. 16, where the available data at 0° and 180° are drawn. The consistency of these summarised data is reasonably good. The next figure (Fig. 17) presents the total cross section extracted from the discussed data and also shows some of the data available at higher energies [24]. Here one of the main differences to the older data shows up in a somewhat higher maximum. A second main difference to the older π^+ data is a more pronounced foreward peak of the differential cross section above the resonance.

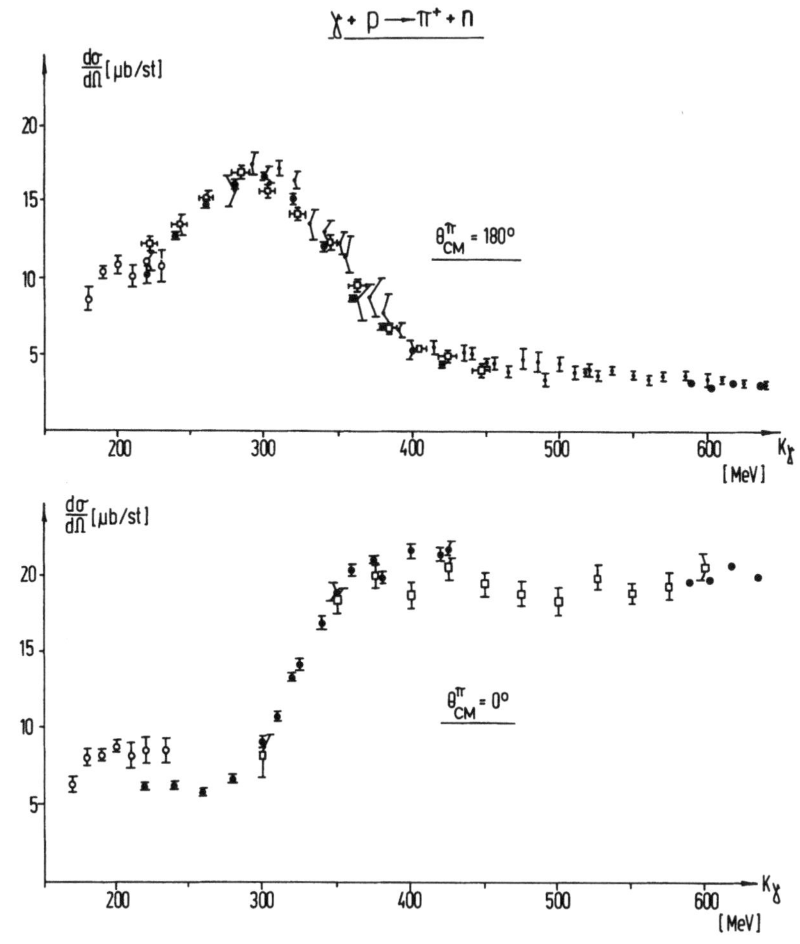

Fig. 16. Comparison of recent π^+ photoproduction data at $\theta^{\pi}_{CM}=0^0$ and $\theta^{\pi}_{CM}=180°$. Moscow, Ref. [16], (fit extrapolations); Bonn, Ref. [18], (fit values); Orsay, $\theta^{\pi}_{CM}=0^0$, Ref. [17a], $\theta^{\pi}_{CM}=180°$, Ref. [17b]; Tokyo, Ref. [21]; Caltech, Ref. [24], (fit values)

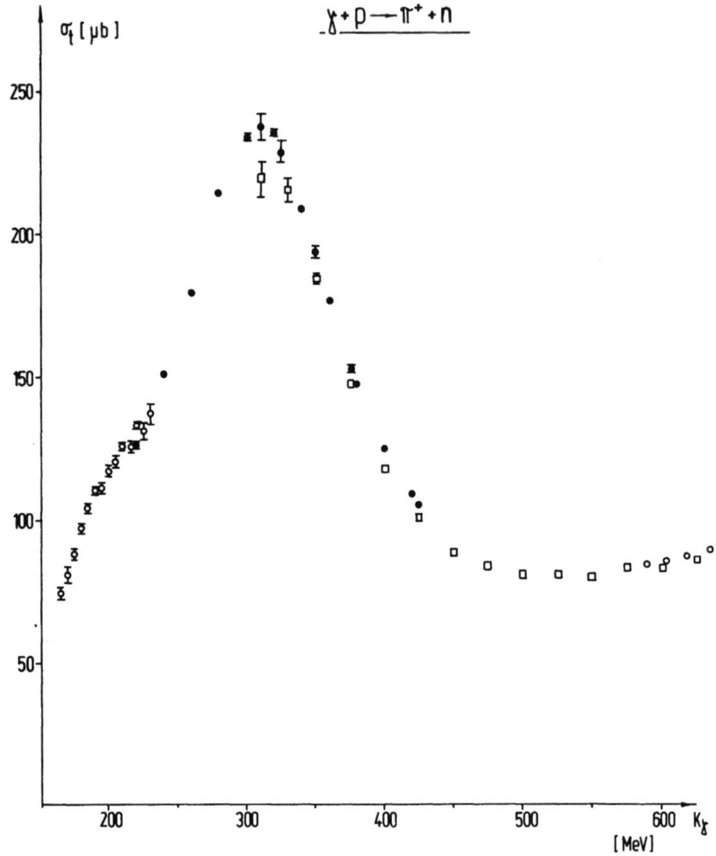

Fig. 17. π^+-photoproduction total cross section. For the point labels see caption of Fig. 15

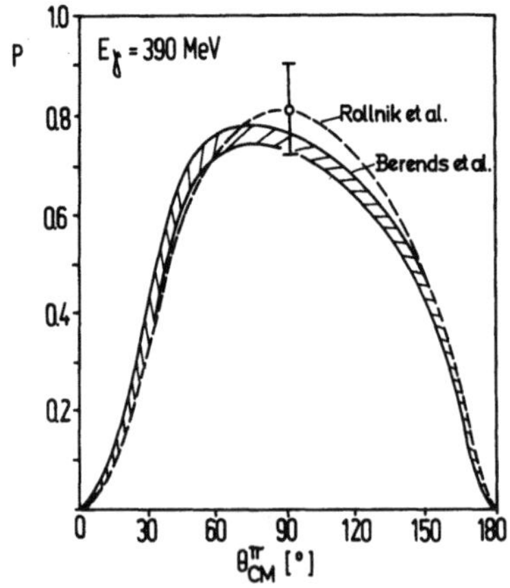

Fig. 18. Polarisation of the recoil neutron from the reaction $\gamma p \to \pi^+ n$ ($E_\gamma = 390$ MeV, $\theta^\pi_{CM} = 90°$) in comparison with theoretical predictions

Coming now to the polarisation of the recoil neutron in the discussed reaction, this quantity has been measured up to now only by one group. In the energy region considered here there is only one point at 390 MeV and 90° (Fig. 18) [25]. This measurement confirmed the rather high neutron polarization, which was predicted by several theoretical models in the first resonance region.

The Bonn group [26] measured further points at 500 MeV and higher energies, which will be presented in the talk of *H. Fischer* [40].

IV. Photoproduction of π^- Mesons on Neutrons

The next reaction, I want to discuss, is the photoproduction of negative pions from neutrons. Due to the absence of a neutron target, here the experimental situation is rather poor, at least compared with the up to now discussed reactions.

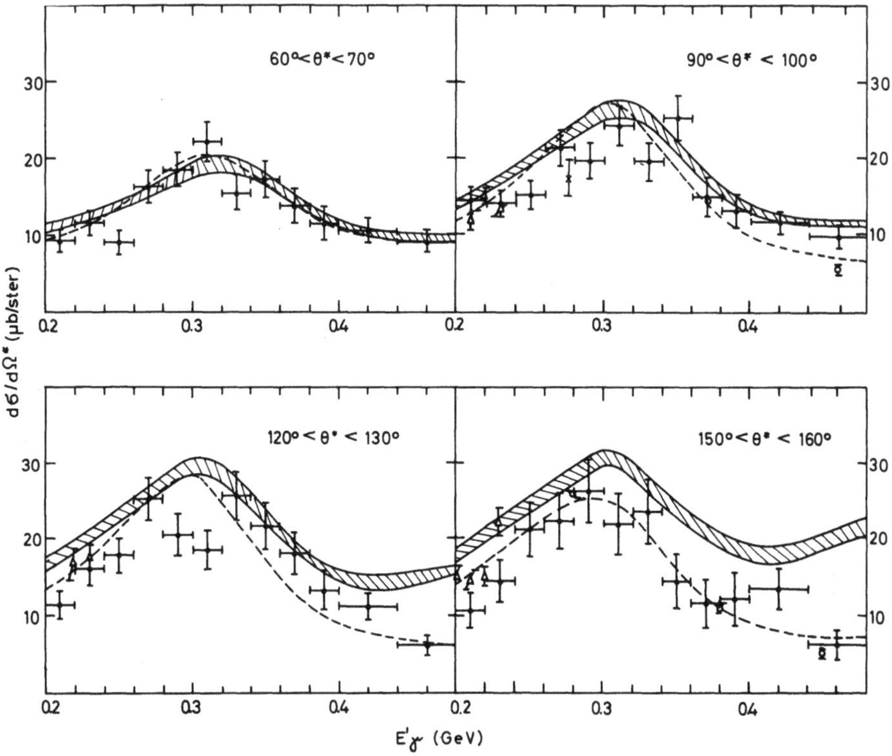

Fig. 19. Differential cross section of the reaction $\gamma n \rightarrow p\pi^-$ as a function of the photon energy in the neutron rest system for four angular intervals. The experimental points are labelled: ABBHHM-collaboration, Ref. [27]; *M. Beneventano et al.*, Ref. [28]; *M. Sands et al.*, Ref. [29]; *D. Garelick et al.*, Ref. [35]

In the considered energy region there are three groups of different measurements available.

At first experiments which determine the π^-/π^+-ratio from the deuteron, measuring only angle and momentum of the outcoming pion. Assuming that in this ratio the deuteron-physics cancels, from the knowledge of the π^+ photoproduction cross section on protons it is possible to obtain the π^- cross section on neutron. This method offers a great advantage on account of its simplicity and for the symmetry between the measure of π^- and π^+, but it has the disadvantage of a bad energy resolution caused by the spectrum of momenta of the target nucleon.

The second group of experiments analyses two or even three of the final state particles, using nuclear emulsion or bubble chambers. Here the evaluation of the cross section is based on the detailed application of the spectator model. The many questions, arising from this problems, will be discussed in a special seminar of this symposium [41].

Next some data are presented yielded by the described experimental methods. Fig. 19 shows excitation curves in the resonance region, mainly with data of the German bubble chamber collaboration [27]. Points are also indicated from π^-/π^+-ratio measurements by *Beneventano et al.* [28], *Sands et al.* [29] and *Garelick et al.* [35]. The curves are due to theoretical calculations by *Berends et al.* (full line) [2] and *Schwela* (dashed line) [30].

There is a recent experiment from the Italian deuterium bubble chamber group [31], which measured total and differential cross sections in the resonance region up to 1 GeV. The two bubble chamber experiments agree very well, as can be seen in Fig. 20, giving the total cross section. To compare, also the total cross sections of π^+ and π^0 photoproduction extracted from the recent experiments are indicated.

I should mention, however, that the data of the German collaboration were reanalysed at the moment, using more refined evaluation methods [32]. First results indicate a trend of the cross sections to higher values.

The third type of experiments, applicable to get information about the considered reaction is to measure the inverse process, that is the radiative capture reaction $\pi^- p \to \gamma n$. The π^- photoproduction cross section can be obtained from this by using detailed balance. In this type of experiment the difficulty to detect two neutral particles is increased by the much more probable process of charge exchange pion-nucleon scattering. However, it has the great advantage to get rid of the difficulties connected with the impulse approximation.

The next figure (Fig. 21) shows results of an experiment of this kind, performed with the π^- beam of the CERN cyclotron [33]. They are compared with the π^-/π^+-ratio evaluations and results from the DESY

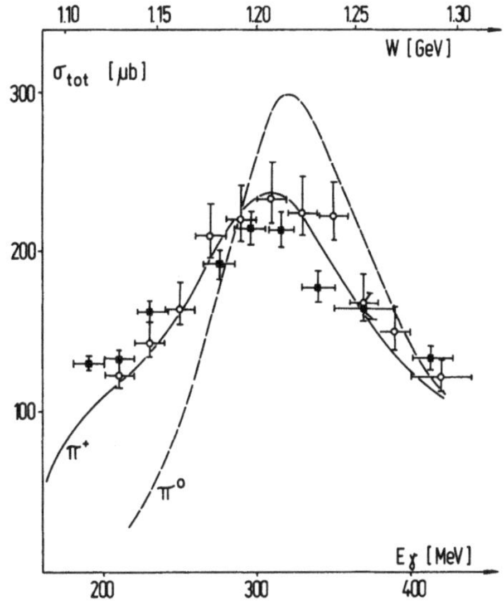

Fig. 20. Total cross sections as a function of the photon energy for the photoproduction reactions: $\gamma n \to \pi^- p$: bubble chamber measurements, ■, Ref. [27], ○, Ref. [31]; $\gamma p \to \pi^+ n$: solid line; $\gamma p \to \pi^0 p$: dashed line

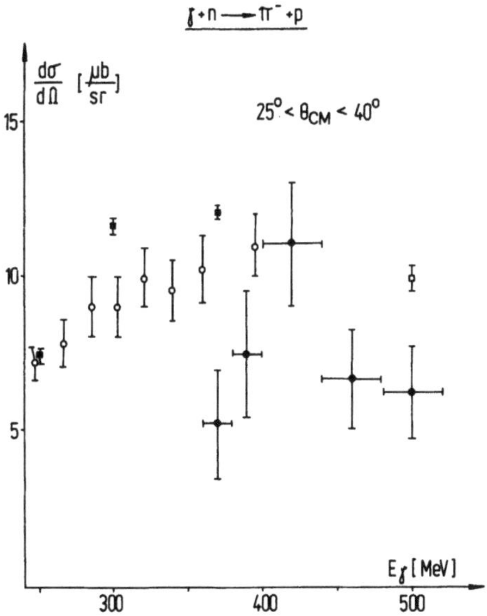

Fig. 21. Comparison of the differential cross section of π^- photoproduction as obtained with different experimental methods: (i) Inverse reaction $\pi^- p \to \gamma n$: ○ *Favier et al.*, Ref. [33]; (ii) π^-/π^+-ratio measurements: ■ *M. Sands et al.*, Ref. [29], □ *G. Neugebauer et al.*, Ref. [34]; (iii) bubble chamber measurements: + ABBHHM-collaboration, Ref. [27]

bubble chamber. The agreement with the ratio measurements is reasonable, whereas there are large deviations to the bubble chamber results.

In spite of the experimental obstacles, the radiative capture reaction provides one of the best means to get reliable information on the photoproduction of negative pions on neutrons. On the other hand, a comparison of the experimental data, extracted from deuteron measurements with those from the inverse reaction may offer a reliable test for the limitations of the impulse approximation.

V. Conclusion

Let me briefly summarise. The experimental situation concerning the differential cross section of single pion photoproduction on protons may be summarised with the help of Fig. 22. Here the goodness of the available data is sketched for photon lab. energies, ranging from threshold up to 500 MeV.

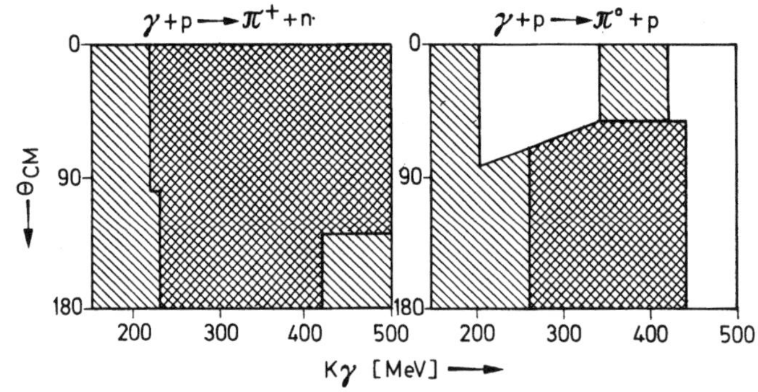

Fig. 22

The notation means the following:

cross hatched area: the measurements in this kinematical region combine to a consistent set of data, with a total uncertainty less than ±10%;

hatched area: the same with a limit of ±20%;

white area: in this region measurements are scarce and uncertain. Further experiments are necessary.

According to Fig. 22 the following conclusions may be given:

$\gamma + p \to \pi^+ + n$. At the moment, there is no need for further measurements of the differential cross section for positive pion photoproduction in the first resonance region. The existing information is sufficient in respect to the precision of the available theoretical models.

$\gamma + p \to \pi^0 + p$. In the case of π^0 photoproduction cross section this no longer holds. There are some white regions, where further accurate measurements are necessary. At first there is the kinematical region of small angles and low energies, i.e. low kinetic energies of the recoil protons, and secondly, the domain above 440 MeV photon lab. energy.

$\gamma + n \to \pi^- + p$. In a similar sketch for π^- photoproduction the whole kinematical region had to be painted white. This is, at the moment, the reaction with the strongest demand for accurate measurements in low energy photoproduction. As was pointed out before, the measurement of the reverse reaction, here, would be of great help.

Concerning the polarisation of the recoil nucleons, there still is a great lack of information, especially in the region of the first resonance.

A further type of valuable information, that was not mentioned up to now can be obtained from experiments with polarised targets. In this field, quite an effort has been done at several laboratories in the last two years. So one can hope to get the first experimental results in the near future.

References

1. *Beale, J. T., Ecklund, S, D., Walker, R. L.:* Report CALT-68-108 (1966).
2. *Berends, F. A., Donnachie, A., Weaver, D.:* Nucl. Phys. B **41**, 103 (1968); Proc. Heidelberg Conf. (1967).
3. *Engels, J., Müllensiefen, A., Schmidt, W.:* Proc. Heidelberg Conf. (1967); – Phys. Rev. **169**, 1796 (1968); – Phys. Rev. **175**, 1951 (1968).
4. *Rollnik, H., Schwela, D., Weizel, R.:* Proc. Heidelberg Conf. (1967). – *Schwela, D., Weizel, R.:* Z. Physik **221**, 71 (1969).
5. *Pfeil, W.:* Thesis, Bonn PI 2–46 (1968). – *Kim Yong Taik:* Report Bonn, PI 2–51 (1968).
6. *Walker, R. L.:* Phys. Rev. **182**, 1729 (1969).
7. *Noelle, P., Pfeil, W., Schwela, D.:* Report Bonn PI 2–79 (1970).
8. *Fischer, G., Fischer, H., von Holtey, G., Kämpgen, H., Knop, G., Schulz, P., Wessels, H., Braunschweig, W., Genzel, H., Wedemeyer, R.:* Report Bonn, PI 1–44 (1968); – Nucl. Phys. B **16**, 93 (1970).
9. *Morand, R., Erickson, E. F., Pahin, J. P., Croissiaux, M. G.:* Phys. Rev. **180**, 1299 (1969).
10. *Hilger, E., Roegler, H. J., Simons, L. M., Tonutti, M.:* Report Bonn, PI 1–103 (1970).
11. *Kemen, H., Wedemeyer, R., Weyer, H. J.:* in progress.

12a. *Hitzeroth, W.:* Nuovo Cimento **60** A, 467 (1969).

12b. *Govorkov, B. B., Denisov, S. P., Minarik, E. V.:* Sov. J. Nucl. Phys. **6**, 370 (1968).

12c. *Miller, D. B., Bellamy, E. H.:* Proc. Phys. Soc. **81**, 343 (1963).

12d. *Koester, L. J., Mills, F. E.:* Phys. Rev. **105**, 1900 (1957).

13. *Althoff, K. H., Kramp, K., Matthäy, K., Piel, H.:* Z. Physik **194**, 135 (1966); – Z. Physik **194**, 144 (1966). – *Feller, P., Niehaus, W., Minatti, N., Piel, H. Trines, D., Unger, M.:* Int. Conf. Electron and Photon Interaction Liverpool (1969); XV. Int. Conf. on High Energy Physics Kiew (1970).
14. *Lunquist, D. E.:* Phys. Rev. **188**, 1527 (1968).
15. *Hayakawa, S., et al.:* J. Phys. Soc. Japan **25**, 307 (1968).

16. *Adamovich, M. I., Larinova, U. G., Latypova, R. A., Lebedev, A. F., Karlamov, S. P., Yagudina, F. R.:* Sov. Phys. Doklady **9**, 792 (1965). – *Adamovich, M. I., et al.:* Sov. J. Nucl. Phys. **7**, 579 (1968).
17a. *Bétourné. C., Bizot, J. C., Perez-y-Jorba, J., Treille, D., Schmidt, W.:* Phys. Rev. **172**, 1343 (1968).
17b. *Pahin, J. P., Erickson, E. F., Morand, R., Croissiaux. M.:* Contr. to the Int. Symp. on Electron and Photon Interactions at High Energies, Liverpool (1969).
18. *Fischer, G., Fischer, H., Heuel, M., von Holtey, G., Knop, G., Stümpfig, J.:* Nucl. Phys. B **16**, 119 (1970); Report Bonn, PI 1–101; Contr. to the XV Int. Conf. on High Energy Physics, Kiew (1970).
19. *Alexandrov, Y. M., et al.:* Phys. Letters **25** B, 372 (1967). – Nucl. Phys. B **10**, 145 (1969).
20. *Alvarez, R. A.:* Phys. Rev. **142**, 957 (1966).
21. *Fujii, T., et al.:* XV. Int. Conf. High Energy Physics, Kiew (1970).
22. *Freitag, D., Schwille, W., Wedemeyer, R.:* Z. Physik **186**, 1 (1965).
23. *von Holtey, G., Stümpfig, J., Wedemeyer, R.:* in progress.
24. *Thiessen, H. A.:* Ph. D. Thesis, Caltech (1966).
25. *Althoff, K. H., Piel, H., Wallraff, W., Wessels, G.:* Phys. Letters **26** B, 640 (1968).
26. *Hahn, U., Wallraff, W.:* Report Bonn, PI 1–087 (1970).
27. ABBHHM-Collaboration: Nucl. Phys. B **8**, 535 (1968).
28. *Beneventano, M., Finzi, R., Mezzetti, L., Paduzi, L., Tazzari, S.:* Nuovo Cimento **10**, 1109 (1958).
29. *Sands, M., Teasdale, J. G., Walker, R. L.:* Phys. Rev. **95**, 592 (1954).
30. *Schwela, D.:* Z. Physik **221**, 158 (1969).
31. *Carbonara, F., et al.:* Report Frascati, LNF–70/16 (1970).
32. *Nellen, B.:* private communication.
33. *Favier, J., et al.:* CERN Preprint 1970.
34. *Neugebauer, G., Wales, W., Walker, R. L.:* Phys. Rev. **119**, 1726 (1960).
35. *Garelick, D., Cooperstein, G.:* Phys. Rev. **136** B, 201 (1964).
36. *Walker, R. L.:* Int. Symp. on Electron and Photon Interactions at High Energies, Liverpool (1969).
37. *Burq, J. P.:* Ann Phys. (Paris) **10**, 363 (1965).
38. *Pecci, R. D.:* Phys. Rev. **181**, 1902 (1969).
39. *Bhatia, M. S., Narayanaswany, P.:* Phys. Rev. **172**, 1742 (1968).
40. *Fischer, H.:* Talk given at this Symposium.
41. *Müller, K.:* Talk given at this Symposium.
42. *Bologna, G.:* To be published in Springer Tracts.
43. *Berends, F. A., Weaver, D. L.:* MIT preprint, Nov. 1970.

Dr. *G. von Holtey*
Physikalisches Institut der Universität Bonn
5300 Bonn, Germany

Pion Photoproduction in the Region of the Δ (1230) Resonance*

D. SCHWELA

Contents

I. Introduction and Motivation

Much work has been done since the appearance of the basic considerations of *Chew*, *Goldberger*, *Low* and *Nambu* [1] in 1957, in order to understand better the single pion photoproduction on nucleons in the region of the first pion nucleon resonance. However, in spite of the most elaborate and sophisticated calculations [2–7] the various theoretical approaches do not agree with *all* experimentally observed quantities quantitatively. In particular, looking at the results of the approach of Refs. [2], which appear to represent the data better than all other calculations, we observe:

a) discrepancies between theoretical predictions or fit values [2] exist for the differential cross sections [8] for $\gamma p \to p\pi^0$ at pion CM angles θ below $\theta = 80°$ and photon lab. energies between $E_\gamma = 160$ MeV and $E_\gamma = 330$ MeV.

b) the recoil proton polarization for $\gamma p \to p\pi^0$ at large pion angles appears to be larger than the experimental results [9],

c) the prediction of the asymmetry ratio Σ (90)° for polarized photons at $\theta = 90°$ lies much below the experimental data [10, 10a] at energies below $E_\gamma = 300$ MeV and lies above experimental data [10, 10a, 10b] above $E_\gamma = 380$ MeV;

d) for $\gamma p \to n\pi^+$ the new measurements [11, 12] of the differential cross sections above $E_\gamma = 380$ MeV are somewhat higher than the theoretical predictions.

---

* This talk presents the work of *P. Nölle*, *W. Pfeil* and *D. Schwela:* "Multipole Analysis of pi-zero and pi-plus photoproduction between threshold and 500 MeV".

This situation is quite similar for the other approaches [3–6], in which, however, even larger deviations for the differential cross sections of $\gamma p \to p\pi^0$ occur (see in particular Refs. [3, 4, 6]). In Ref. [7] the prediction of the recoil proton polarization disagrees completely with the data [9], moreover, the approach is not able to describe π^0- and π^+-photoproduction at the same time (with the same coupling parameters).

All theoretical work has confirmed the result of CGLN [1] i.e. the predominance of the magnetic dipole $M_{1+}^{3/2}$ and the fact that $E_0^{\pi+} = E_0^{\pi+}$ (Born). The results for the other multipoles differ quite substantially in the different papers and the situation for these "small" multipoles is still not clear.

It is not easy to decide which multipole amplitude is responsible for the discrepancies one observes between theory and experiment. Therefore it is useful to make an energy independent phenomenological multipole analysis, which may indicate where the theoretical approaches are wrong. Attempts at this goal have been made in the past years [13–20] but the results are not very persuasive as in some of them [16, 17] the investigations are restricted to small energy regions and/or to π^+ photoproduction only [15–16, 20] or are dependent on assumptions which are not on a sure footing [13–15, 19], but which had to be made because of lack of accurate and consistent data.

Fortunately, with the advent of the new and very accurate data of Bonn [8, 11] and Orsay [12, 21] this problem is overcome, and the present data of π^0 and π^+ photoproduction allow a sensitive analysis [22] with a minimum number of assumptions made.

II. Results of the Multipole Analysis

The data for differential cross sections $d\sigma/d\Omega(\theta)$, asymmetry ratios for polarized photons Σ (θ) and recoil nucleon polarizations $P(\theta)$ of Refs. [8–12, 23–28] have been taken in order to determine in an *energy independent* fit the four multipole amplitudes

$$E_{1+}(W),\ M_{1-}(W),\ E_{1+}(W) \text{ and } M_{1+}(W)$$
$$\text{for } \gamma p \to p\pi^0,\ \gamma p \to n\pi^+ \text{ at 19 energies}$$
$$\text{between } E_\gamma = 180\ \text{MeV and } E_\gamma = 500\ \text{MeV}.$$

The $(l \geqq 2)$ – multipoles have been taken in Born approximation, and the Watson theorem has been used for the $(l \leqq 2)$ multipoles.

When trying a multipole analysis of pion photoproduction in the region of the first pion nucleon resonance one has to be aware of the fact that in this energy region photoproduction is largely determined by the magnetic dipole $M_{1+}^{3/2}$. At those energies, where asymmetry ratios

$\Sigma(\theta)$ or recoil nucleon polarizations $P(\theta)$ are not present or scarce, it is therefore particularly difficult to fix the small multipoles and one has to expect several possible solutions. In fact, in the beginning of the analysis, at each energy several solutions were found, except at $E_\gamma = 360$ MeV where only *one* solution was obtained, this being due to the existence of a fairly complete angular distribution of the recoil proton polarization $P(\theta)$ for $\gamma p \to p\pi^0$.

However,

1. by requiring that a continuous connection should exist between adjacent energies, covering the whole energy region

2. by looking at the detailed values for the polarization $P(\theta)$ and the asymmetry $\Sigma(\theta)$ and not only at an overall χ^2 value,

these solution were reduced to *two sets of multipoles*, phrased by "solution A" and "solution B".

These two solutions are best characterized by looking at the magnetic dipole $M_{1+}^{3/2}$. In Fig. 1 the Argand plot of solution A for this multipole is displayed. All values (note the small error bars!) lie remarkably well on a circle (solid line). This circle is not put in by the Watson theorem!

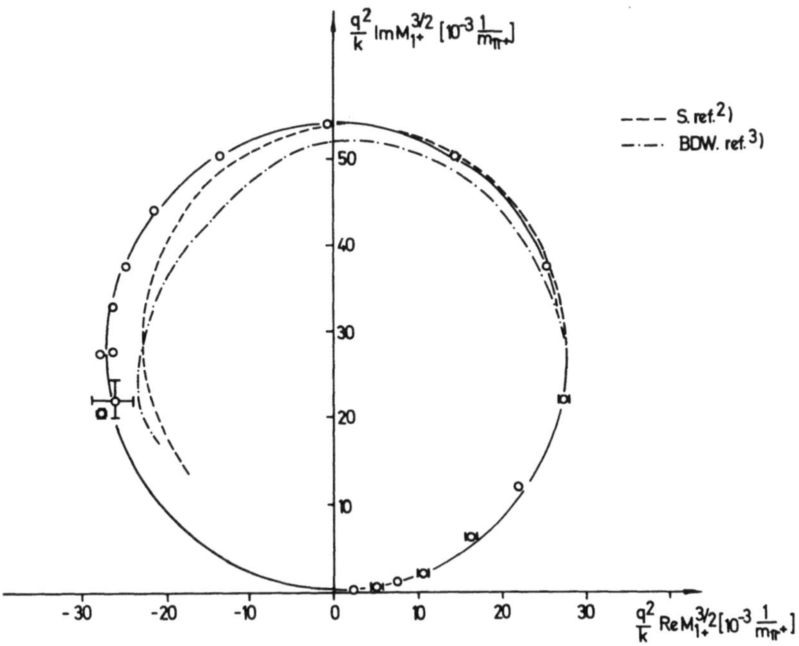

Fig. 1. Argand plot of fit values ○ for $q^2/k\ M_{1+}^{3/2}$, solution A, compared to dispersive calculations [2, 3]. The numbers quoted are photon lab. energies. The solid line circle is an eyeball fit to the fit results

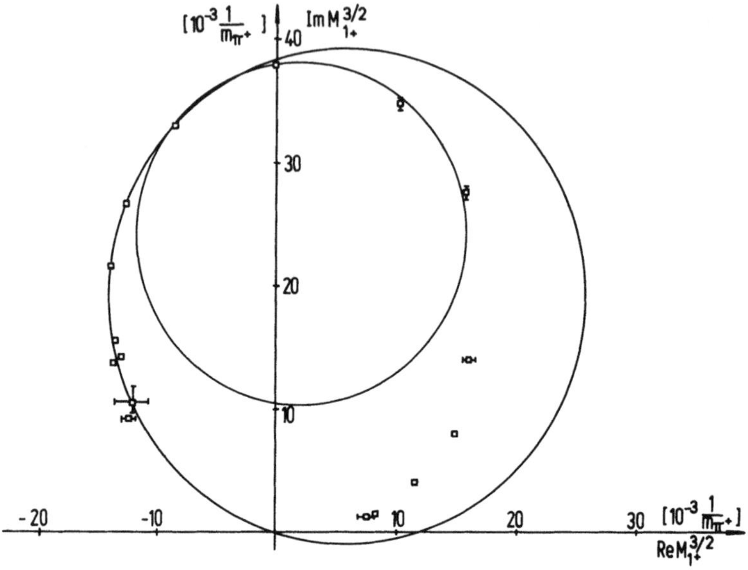

Fig. 2. Argand plot of fit values □ for $M_{1+}^{3/2}$, solution B, compared to solution A

In Fig. 2 the Argand plot of solution B is shown. Clearly the Argand circle is deformed. Above resonance, solution B coincides with solution A. Another feature of solution B is the much stronger energy dependence of the other multipole amplitudes. It should be stressed, that solution B fits the experimental data of $\mathrm{d}\sigma/\mathrm{d}\Omega$, Σ and P as satisfactorily as solution A does. Only a measurement of the polarized target up – down asymmetry

$$T(\theta)=\frac{\dfrac{\mathrm{d}\sigma^{\uparrow}}{\mathrm{d}\Omega}-\dfrac{\mathrm{d}\sigma^{\downarrow}}{\mathrm{d}\Omega}}{2\dfrac{\mathrm{d}\sigma}{\mathrm{d}\Omega}} \tag{1}$$

for π^0- and π^+-production *at small pion CM* angles and *below resonance* could decide between solution A and B as the following table shows for $T(60°)$ at $E_\gamma=240$ and 300 MeV.

sol. / E_γ	A	B	A	B
240	35.1	24.0	27.2	14.1
300	12.9	3.1	71.1	49.1
	$\gamma p\to p\pi^0$		$\gamma p\to n p\pi^+$	

The theoretical explanation of the Argand plot of Fig. 2 as a $\Delta(J^P=\frac{3}{2}^+, I=\frac{3}{2})$ – resonance is difficult as the total width of the resonance appears to be

$$\Gamma^B_{tot}=70\ \text{MeV} \tag{2}$$

while from solution A

$$\Gamma^A_{tot}=110\ \text{MeV} \tag{3}$$

as one expects from πN scattering results for P_{33}. For this reason, in the following only solution A will be discussed in some more detail.

Let us now compare the results for $M^{3/2}_{1+}$ to the results of two dispersive calculations [2, 3] which are the most elaborate and sophisticated ones. From Fig. 1, one observes in the below resonance region deviations up to 10% above some larger discrepancies occur between the fit result and the dashed (S., Ref. [2]) and dash-dotted (BDW., Ref. [3]) curves, respectively. (It should be noted that the equality of the Argand curves of Refs. [2, 3] outside the resonance region is apparent, as the energy scales are different.) On the scale of $M^{3/2}_{1+}$ these deviations are quite substantial. However, they might be trivially caused by the use of different phase shifts in Refs. [2] and [3] as compared to the P_{33} phase [29] used in the multipole analysis.

In Fig. 3 the result of the analysis for the magnetic dipole M_{1-} is presented. In both the isospin 1/2 and 3/2 amplitudes the errors are rather large (except at threshold and $E_\gamma=450{,}475$ MeV, where some parameters had to be kept fixed at reasonable values), indicating that accurate measurements of angular distributions of the asymmetry ratio $\Sigma(\theta)$ for $\gamma p\to p\pi^0$ are required.

The isospin 1/2 part of M_{1-} is in general agreement with theoretical results. The fit values for $M^{3/2}_{1-}$, however, are absolutely smaller by a factor of approximately 2, which explains the discrepancies b) and c) quoted in the introduction. This is somewhat surprising as everybody would expect dispersion theory not to be correct rather in case of the isospin 1/2 part (as the phase P_{11} involved is inelastic above $E_\gamma=400$ MeV) than in case of the isospin 3/2 part (as the phase P_{31} involved is practically elastic up to $E_\gamma=800$ MeV). A possible explanation could be the inclusion of ω-exchange, which is known to change $M^{3/2}_{1-}$ substantially (cf. Ref. [30]). Unfortunately, when adding ω-exchange in the usual manner, cf. Refs. [2, 3] one runs the risk of double counting [2].

Fig. 4 shows the electric dipole $E^{3/2}_{0+}$. Below resonance the error bars are larger than above. This corresponds to the scarcity of differential cross section data for $\gamma p\to p\pi^0$ at pion CM angles θ below $\theta=70°$.

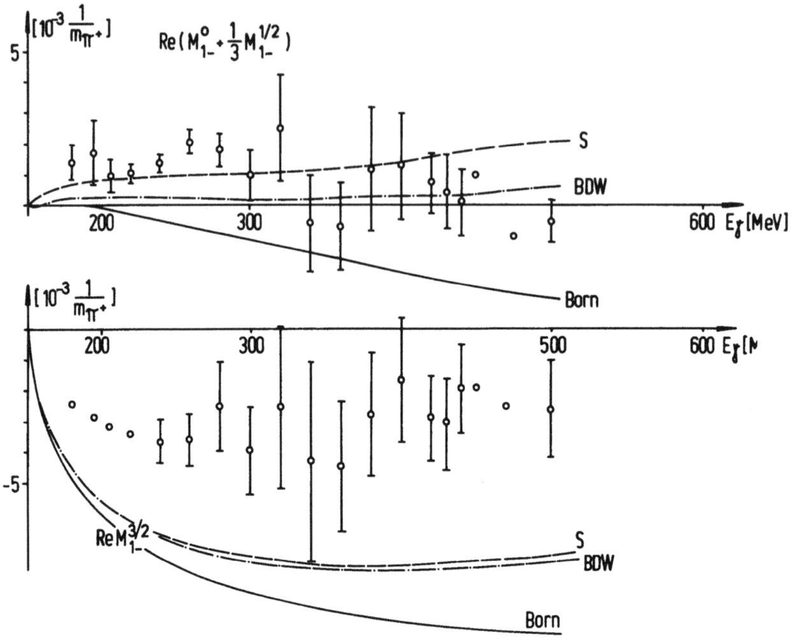

Fig. 3. Isotopic parts of fitted ReM_{1-}, compared to Born approximations and results of Refs. [2, 3]

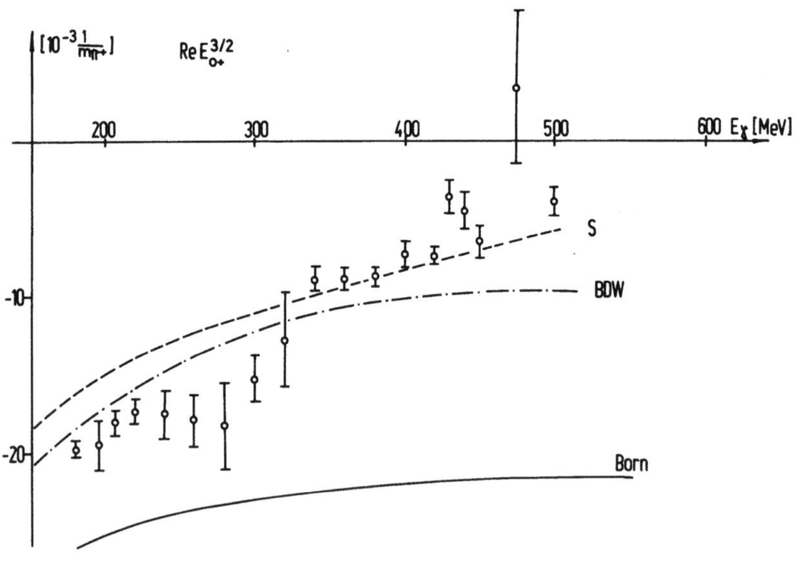

Fig. 4. Real part of fitted $E^{3/2}_{0+}$, compared to Born approximation and results of Refs. [2, 3]

The energy dependence of this multipole seems to be somewhat steeper than that predicted by the dispersive calculations.

In Fig. 5 the ratio

$$E_{1+}^{3/2}/M_{1+}^{3/2}$$

is shown. Some evidence for a zero of $E_{1+}^{3/2}$ near $E_\gamma = 360$ MeV exists indicating a CDD-zero in this amplitude as has been assumed in the

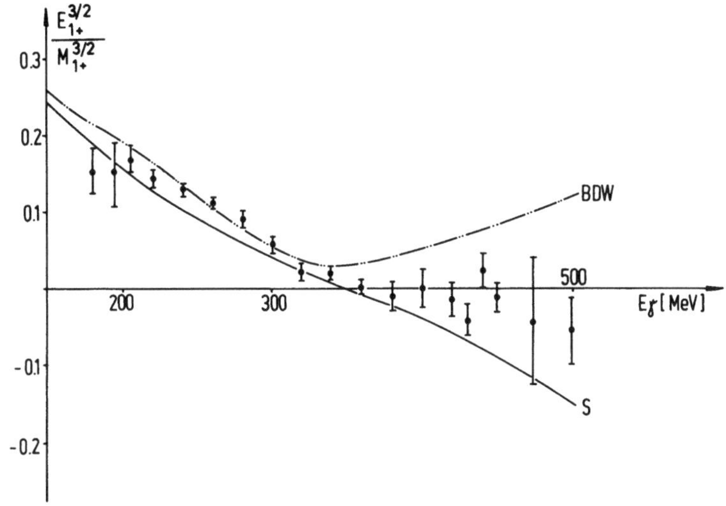

Fig. 5. The fit-values of the ratio $E_{1+}^{3/2}/M_{1+}^{3/2}$ compared to the results of Refs. [2, 3]

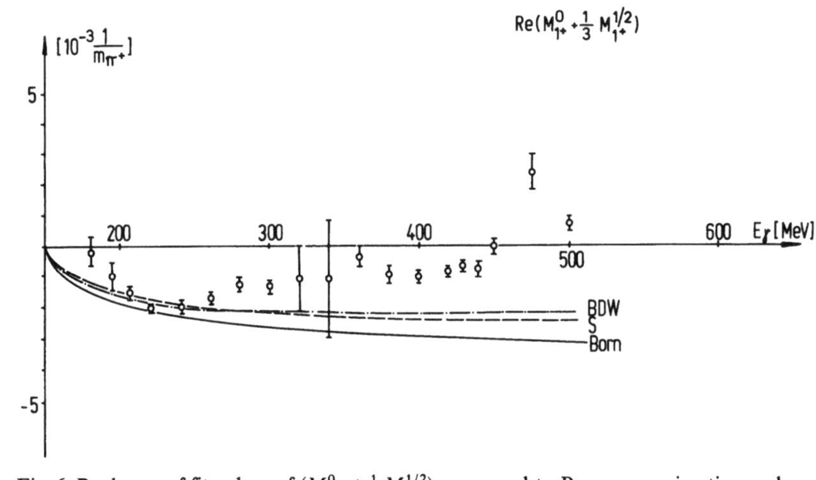

Fig. 6. Real part of fit values of $(M_{1+}^{0} + \frac{1}{3} M^{1/2})$ compared to Born approximation and results of Refs. [2, 3]

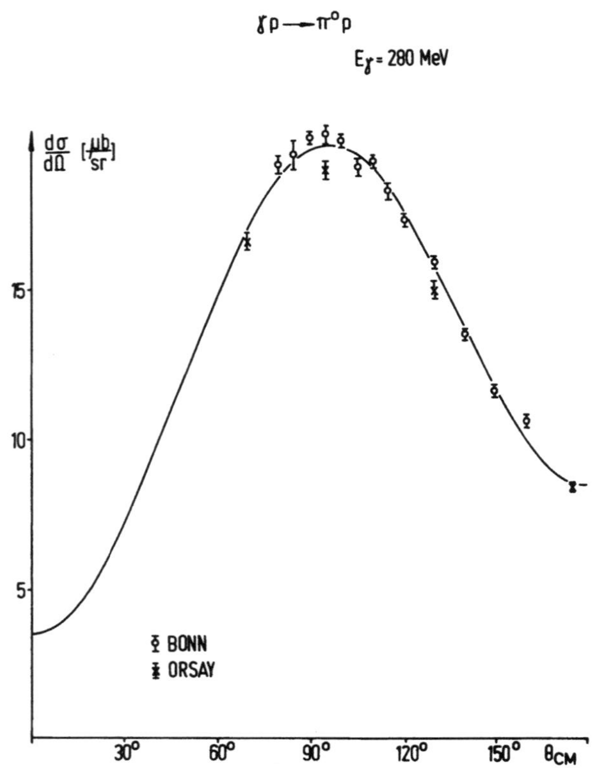

Fig. 7. Angular distribution of differential cross section at $E_\gamma = 280$ MeV for $\gamma p \to p\pi^0$. The solid line is the result of the analysis. Data are taken from Refs. [8, 21]

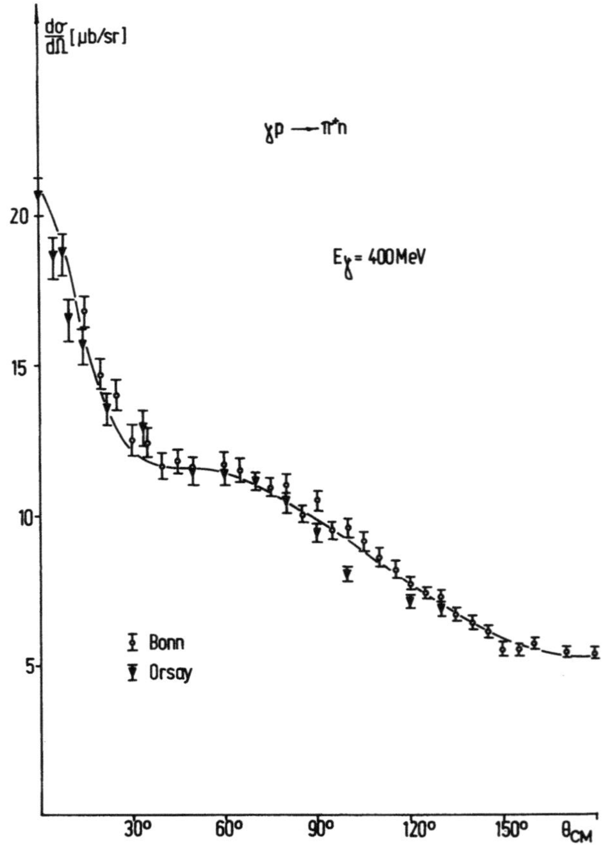

Fig. 8. Same as Fig. 7 at $E_\gamma = 400$ MeV for $\gamma p \to n\pi^+$. Data are taken from Refs. [11, 12]

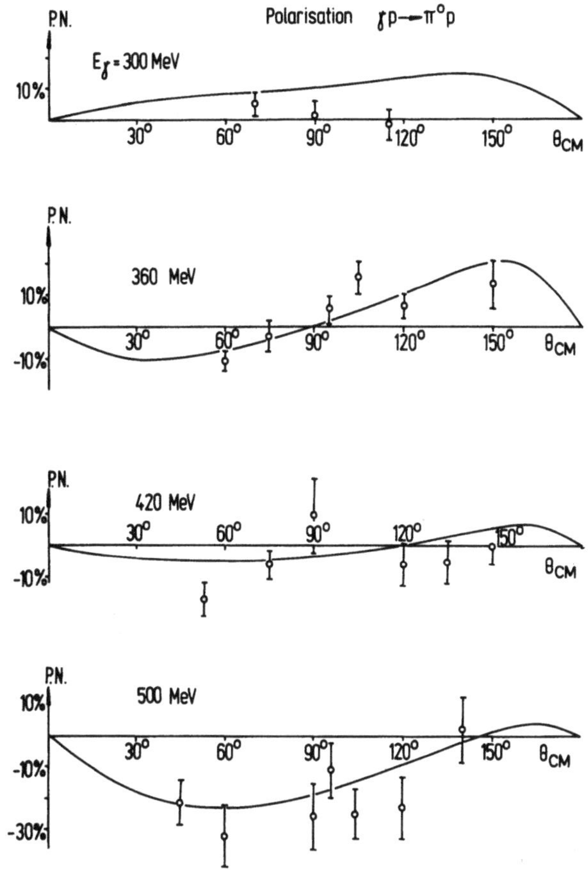

Fig. 9. Recoil proton polarization at $E_\gamma = 300$, 360, 420 and 500 MeV. Data are taken from Ref. [9]

theoretical literature [2, 4] Above $E_\gamma = 360$ MeV, the result of BDW deviates largely while that of Ref. [2] gives the correct slope.

As an example of a really small amplitude in Fig. 6 the real part of $M_{1+}^{0} + \frac{1}{3} M_{1+}^{1/2}$ is shown. This amplitude, too, is well determined, except in the immediate resonance region. The fit result seems to exhibit a zero near $E_\gamma = 450$ MeV, which is not present in the theoretical calculations. A very similar situation also occurs for $\mathrm{Re}(E_{1+}^{0} + \frac{1}{3} E_{1+}^{1/2})$.

The experimental data used in the fit are very well fitted, as is illustrated in Fig. 7–10, where differential cross sections for π^0- and π^+-production, recoil nucleon polarizations and asymmetry ratios are shown.

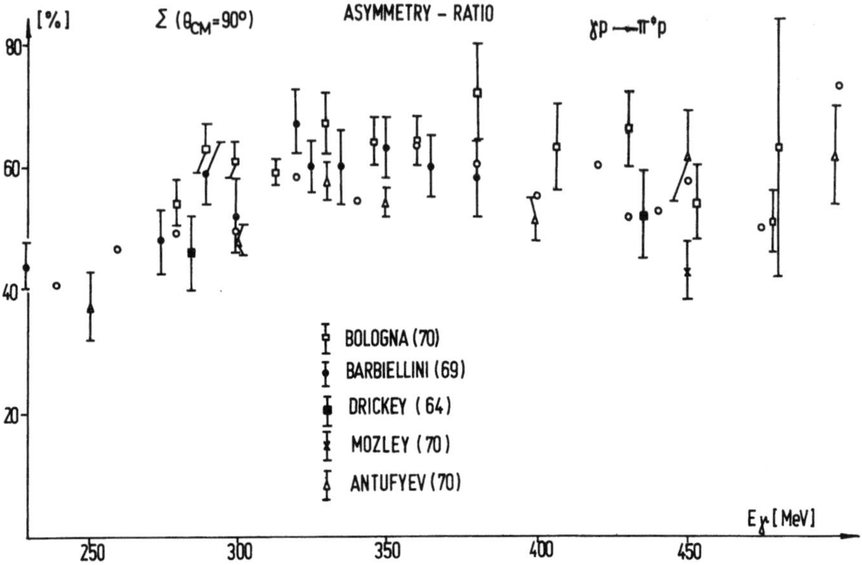

Fig. 10. Excitation curve of asymmetry ratio Σ (90)° for $\gamma p \to p\pi^0$. Open circles are the result of the analysis. Data are taken from Refs. [10, 10a, 10b, 10c, 25]

III. Conclusions

1. A consistent and satisfactory determination of the ($l \leqq 1$)-multipole amplitudes of single pion photoproduction in the region of the first pion nucleon resonance has been obtained. Disregarding small jumps at resonance the energy dependence of the multipoles obtained is smooth. All data included in the analysis are well represented.

2. The magnetic dipole amplitude $M_{1+}^{3/2}$ shows impressively the Breit Wigner shape of the P_{33} resonance in photoproduction.

3. The magnetic dipole $M_{1-}^{3/2}$ predicted by dispersive calculation turns out to be wrong by a factor of 2.

4. In order to reduce the error bars of various multipoles – in particular M_{1-} and E_{0+} below resonance – one should measure

$$\underline{\frac{d\sigma}{d\Omega}(\gamma p \to p\pi^0) \quad \text{for} \quad \theta_{CM}^{\pi} \leqq 80°,\ E_\gamma \leqq 340\ \text{MeV},}$$

angular distributions of the *asymmetry ratio* $\Sigma(\theta)$ and the *recoil proton polarization* $P(\theta)$ for $\gamma p \to p\pi^0$.

In order to be able to distinguish between solution A and B the polarized target quality $T(\theta)$ defined in Eq. (1) should be measured for $\theta_{CM}^{\pi} \lesssim \frac{\pi}{2}$.

Note added in proof: In the meantime an analysis similar in mind to ours has been done by *Berends* and *Weaver* [31]. They reproduce essentially the main results of this work, in particular that for $M_{1-}^{3/2}$. Some discrepancy exists between their $M_{1+}^{3/2}$ and ours above resonance. This fact, however, may be due to their use of the old P_{33} CERN phase shift [32], which differs quite substantially from the new one of *Bugg et al.* [29] used in our work.

References

1. *Chew, G. F., Goldberger, M. L., Low, F. E., Nambu, Y.*: Phys. Rev. **106**, 1345 (1957) (referred to as CGLN).
2. *Schwela, D., Rollnik, H., Weizel, R., Korth, W.*: Z. Physik **202**, 451 (1967); – *Schwela, D., Weizel, R.*: Z. Physik **221**, 71 (1969); – *Schwela, D.*: Technical Report No. 2, Bonn, December 1970 (referred to as S).
3. *Berends, F. A., Donnachie, A., Weaver, D.*: Nucl. Phys. B **4**, 1, 54, 103 (1969) (referred to as BDW).
4. *Engels, J., Schmidt, W.*: Phys. Rev. **169**, 1296 (1968). – *Engels, J., Müllensiefen, A., Schmidt, W.*: Phys. Rev. **175**, 1951 (1968).
5. *von Gehlen, G.*: Nucl. Phys. B **20**, 102 (1970).
6. *Adler, S. L.*: Ann. Phys. **50**, 189 (1968).
7. *Kim Yong Taik*: Diplomarbeit, Bonn 1968. – *Pfeil, W., Thesis. Ph. D.*: Bonn 1968.
8. *Fischer, G., Fischer, H., von Holtey, G., Kämpgen, H., Knop, G., Schulz, P., Wessels, H., Braunschweig, W., Genzel, H., Wedemeyer, R.*: Nucl. Phys. B **16**, 93 (1970). – *Hitzeroth, W.*: Nuovo Cimento **60**, 467 (1969).
9. *Blüm, P., Brinkmann, P., Brockmann, R., Lütter, P., Mohr, W., Feller, P., Minatti, N., Piel, H., Unger, M., Niehaus, W., Trines, D.*: Contribution to the Kiev Conference 1970.
10. *Barbiellini, G., Bologna, G., Capon, C., de Zorzi, G., Fabbri, F. L., Murtas, G. P., Sette, G., Diambrini, G., de Wire, J.*: Phys. Rev. **184**, 1402 (1969).

10a. *Bologna, G., et al.*: LNF Preprint No. 70–139 (1970).

10b. *Antufyev, Y. P., et al.*: Kharkov Physico-Technical Institute Preprint 1970.

10c. *Mozley, R. F.*: private communication.

11. *Fischer, G., Fischer, H., Heuel, M., von Holtey, G., Knop, G., Stümpfig, J.*: Nucl. Phys. B **16**, 119 (1970).
12. *Bizot, J., Perez-Y-Jorba, J. P., Treille, D.*: Phys. Letters **25** B, 489 (1967).
13. *Berends, F. A., Donnachie, A.*: Contribution No. 140 to the 4th International Symposium on Electron and Photon Interactions at High Energies, Liverpool 1969.
14. *Walker, R. L.*: Phys. Rev. **182**, 1729 (1969).
15. *Grilli, M., Spillantini, P., Valente, V., Nigro, M., Schiavuta, E.*: Nuovo Cimento **68**, 1188 (1969).
16. *Aleksandrov, Y. M., Grushin, V. F., Leikin, E. M.*: Preprint No. 46 of the Lebedev Institute, Moscow (1969).
17. *Müllensiefen, A.*: Z. Physik **211**, 360 (1968).
18. *Kondo, K., Nishikawa, T., Suzuki, T., Takikawa, K.*: J. Phys. Soc. Japan **29**, 30 (1970).
19. *Wennström, T.*: Nucl. Phys. B **5**, 235 (1968).

20. *Spillantini, P., Valente, V., Nigro, M., Oleari, C.:* Nucl. Phys. B **13**, 320 (1969).
21. *Morand, R., Erickson, E. F., Pahin, J. P., Croissiaux, M. G.:* Phys. Rev. **180**, 1299 (1969).
22. *Nölle, P., Pfeil, W., Schwela, D.:* Nucl. Phys. B **26**, 461 (1971).
23. *Govorkov, B. B., Denisov, S. P., Minarek, E. V.:* Sov. J. Nucl. Phys. **6**, 370 (1968).
24. *Hilger, E., Roegler, H., Tonutti, M.:* Contribution to the Kiev Conference, 1970.
25. *Drickey, D. J., Mozley, R. F.:* Phys. Rev. **136**, B 543 (1964).
26. *Grilli, M., Spillantini, P., Soso, F., Nigro, M., Schiavuta, E., Valente, V.:* Nuovo Cimento **54**, 877 (1968).
27. *Blüm, P., et al.:* Contribution to the Liverpool Conference, 1969.
28. Where necessary, some older data have been taken, which are quoted in *Beale, J. T., Ecklund, S. D., Walker, R. L.:* Caltech Report CTSL-42, 1966.
29. *Bugg, D., Bussey, P., Dance, D., Carter, A. A., Williams, J.:* Rutherford Laboratory Preprint RPP/H/70.
30. See *Rollnik, H.:* Proceedings of the Heidelberg International Conference on Elementary Particles, p. 400, edited by H. Filthut. Amsterdam: North Holland publishing Comp. 1968.
31. *Berends, F. A., Weaver, D. L.:* MIT preprint, Nov. 1970.
32. Particle Data Group UCRL-20030, 1970.

Dr. *D. Schwela*
Physikalisches Institut der Universität Bonn
5300 Bonn, Germany

Multiple Pion Photoproduction in the *s* Channel Resonance Region

D. Lüke and P. Söding

Contents

1. Introduction

At all photon energies above the $\Delta(1236)$ region, the total photoproduction cross section is found to exceed the sum of the wellmeasured single meson production cross sections by a large amount (Fig. 1). What kind of processes are responsible for the rest of the cross section? Can we learn anything from them?

To review these processes in the low energy region is somewhat dissatisfying because in multibody photoproduction, the center of interest clearly lies with the *high* energies. In fact most of the results which we will discuss have been obtained in experiments devised to study higher energies, as an unwelcomed by-product due to the use of bremsstrahlung photon beams. The precision of the multiple production data is inferior compared with single meson photoproduction [1]. The data do have interesting features but nothing spectacular has so far been learned from them.

After a brief survey of cross sections we will concentrate on the results and interpretation of the reaction $\gamma p \to \pi^+ \pi^- p$ which is the only one studied in more detail. Other reactions will also be briefly discussed.

2. Cross Sections for Multiple Pion Photoproduction

The total cross section σ_{tot} for photoproduction on protons as a function of energy is shown in Fig. 1 [2]. The bulk of the data, in particular below 1 GeV, is from measurements of inelastic electron scattering by the

SLAC-MIT Group [3], extrapolated to virtual photon mass $q^2 \to 0$. These results are still preliminary. In addition there are direct measurements with real photons, using counters for the reaction products and tagged photon beams [4, 5]; or using hydrogen bubble chambers to detect the reaction products, with either tagged [6] or monochromatic [7, 8] photon beams.

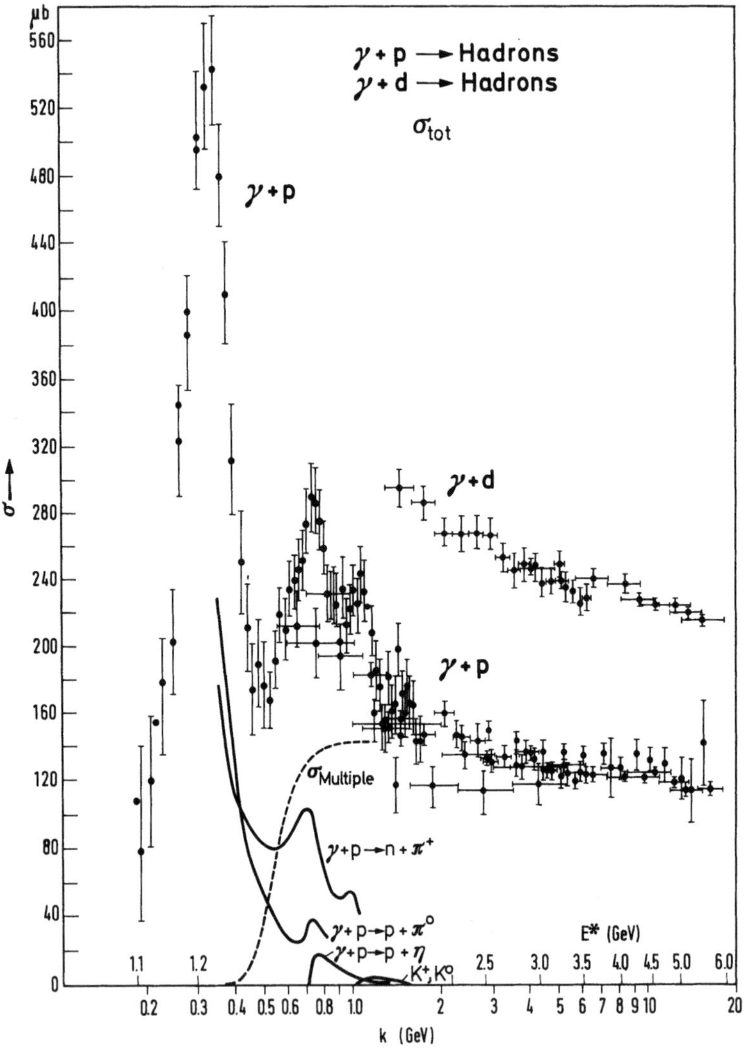

Fig. 1. Total cross sections of photoproduction as a function of laboratory photon energy (k) and total c.m.s. energy (E^*). (Single and multiple meson photoproduction cross sections are shown for comparison.)

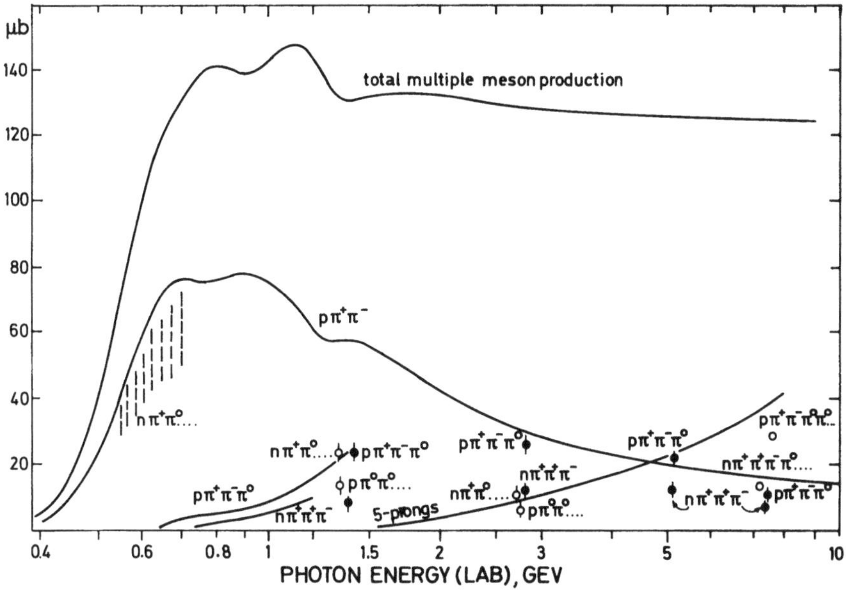

Fig. 2. Cross sections of multiple pion photoproduction

Also shown in Fig. 1 are the various single meson photoproduction cross sections. If one subtracts these from σ_{tot}, one gets a rough idea of the behaviour of the multiple meson photoproduction cross section, as schematically shown in Fig. 1. It rises very rapidly even *below* the region of the second resonance bump in σ_{tot}, reaches a value of $\sim 140\,\mu b$ at 800 MeV and, after some possible wiggles, becomes smooth and equal to the total cross section of about $120\,\mu b$ at high energies.

A breakdown of the multiple meson photoproduction cross section into the more important channels [8–14] is shown in Fig. 2. Below 1.5 GeV the reaction

$$\gamma p \to \pi^+ \pi^- p \tag{1}$$

is dominant. Above 3 GeV, final states involving more than one neutral pion become very abundant, so that only a small fraction of the multiple production cross section has been studied in detail.

3. Measurements and General Results on the Reaction $\gamma p \to \pi^+ \pi^- p$

Of the three possible double-pion photoprocesses on the proton

$$\gamma p \to \pi^+ \pi^- p \tag{1}$$

$$\to \pi^0 \pi^0 p \tag{2}$$

$$\to \pi^+ \pi^0 n \tag{3}$$

only the first one has been studied, since the other two are "unmeasurable" due to the presence of more than one neutral particle in the final state. Of the charge-symmetric reactions

$$\gamma n \to \pi^- \pi^+ n \tag{4}$$

$$\to \pi^0 \pi^0 n \tag{5}$$

$$\to \pi^- \pi^0 p \tag{6}$$

only reaction (5) is "unmeasurable", while cross sections for reactions (4) and (6) have been deduced from measurements of photoproduction on deuterons. We will comment on these in section 5 (see also Table 3 and Fig. 5).

The experiments on the reaction $\gamma p \to \pi^+ \pi^- p$ have been of three basically different types.

The first type is the missing-mass experiment. One measures with counters the momentum and angle of the π^-, using a magnetic spectrometer. With the method of bremsstrahlung edge subtraction one determines the effective photon energy and then calculates the missing-mass in the reaction

$$\gamma p \to \pi^- (\mathrm{MM})^{++}.$$

One finds a distribution with a maximum in the $\Delta^{++}(1236)$ region plus some smooth "background". To the background a number of different processes can contribute. It may be partly reflection from ϱ^0 production in reaction (1), or from reaction (1) without any final-state resonances; but it may also be due to reactions different from reaction (1), such as

$$\gamma p \to \pi^+ \pi^- \pi^0 p \tag{7}$$

$$\to \pi^+ \pi^- \pi^+ n. \tag{8}$$

In an experiment of this type one therefore does not separate reaction (1) from other reactions. One has to make an *assumption* on the behaviour of the background, and can then subtract the background under the $\Delta^{++}(1236)$ bump to get a differential cross section $\mathrm{d}\sigma/\mathrm{d}\Omega$ for the process of $\Delta^{++}(1236)$ production,

$$\gamma p \to \pi^- \Delta^{++}(1236). \tag{9}$$

In this simple type of experiment, one does not obtain any information on the decay of the Δ^{++}. Table 1 gives a list of the more recent experiments performed in this way. An example [16] of the angular distributions obtained for reaction (9) is shown in Fig. 3.

In another type of experiment, one has detected, with spark chambers and/or counters, two (or three) of the three final particles of reaction (1) in coincidence, but without measuring all the momenta and without 4π solid angle detection. A list of such experiments is given in Table 2.

Table 1. *Missing mass experiments on the reaction $\gamma p \to \pi^- \Delta^{++}$*

Author(s)	year	lab.	E_γ (LAB) (GeV)	θ_{CM}	Ref.
Beneventano et al.	65	Frascati	0.60–0.85	≈60°	15
Allaby, Lynch, Ritson	66	Stanford	0.58–0.95	≈all	16
Hauser	67	Caltech	0.90–1.30	≈all	17
Burg et al.	68	Orsay	0.85–1.21	≈90°	18
Morrison, Drickey, Mozley	68	Stanford	0.57–0.80 (polarized)	45° 90°	19

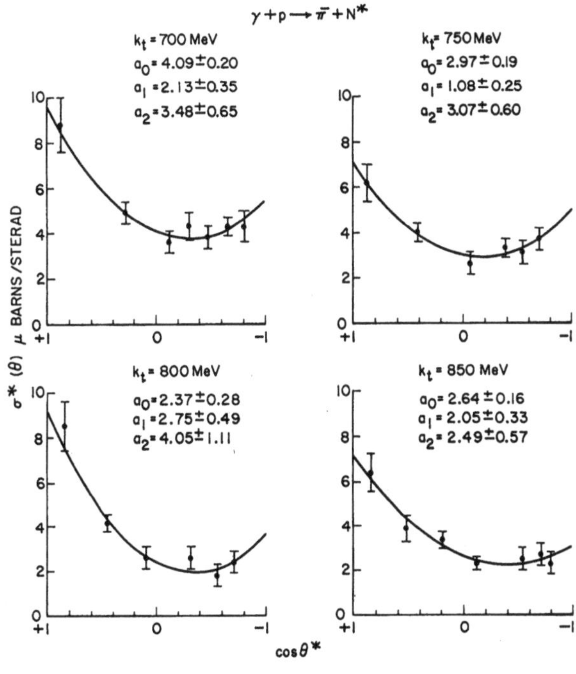

Fig. 3. C.m.s. production angular distributions of the reaction $\gamma p \to \pi^- \Delta^{++}(1236)$ at photon laboratory energies of 700–850 MeV from Ref. [16]. Curves are fits of the form $\sigma^*(\theta) = \Sigma\, a_n (\cos\theta^*)^n$.

For some of these experiments, the interpretation of the results is difficult due to the limited kinematical region which is observed, for example in the production or decay angles of the resonant $\pi^+ p$ system, or in the effective mass $m_{\pi\pi}$ of the two pions. Reflections from other resonances like $\Delta^0 \to \pi^- p$ are then often not well determined, resulting in uncertainty with the background subtraction under the resonances studied (in most cases $\Delta^{++} \to \pi^+ p$, or $\varrho^0 \to \pi^+ \pi^-$).

Table 2. *Coincidence experiments on the reaction $\gamma p \to \pi^+ \pi^- p$ (spark chambers and/or counters)*

Author(s)	year	lab.	E_γ(LAB) (GeV)	number of events	Ref.
Del Fabro et al.	65	Frascati	<1.1	4900	20
Diambrini-Palazzi et al.	68	Frascati	0.55–0.85	7400	21
Ukai et al.	67	Tokyo	<1	2400	22
Shanks	68	Caltech	0.8–1.5	3020	23
Franklin et al.	64	Cornell	0.9–1.3	17000	24
Eisenhandler et al.	68	Cornell	1–2		25
Criegee et al.	70	DESY	1.4–3.3 (polarized)	31000	26

Table 3. *Bubble or streamer chamber experiments*

Group	year	E_γ(LAB) (GeV)	reaction	number of events	Ref.
Cambridge Bubble Chamber Coll.	67	<6	$\gamma p \to p\pi^+\pi^-$	4400	27
ABBHHM Collaboration	68	<6	$\gamma p \to p\pi^+\pi^-$	18800	10
SLAC Streamer Chamber Group	69	<16	$\gamma p \to p\pi^+\pi^-$	2000	28
SLAC-Berkeley-Tufts (Laser)	69	2.8, 4.7 (pol)	$\gamma p \to p\pi^+\pi^-$	5760	29
Frascati-Rome-Pavia-Naples	69	<1	$\gamma p \to p\pi^+\pi^-$	2860	14
Frascati-Rome-Pavia-Naples	70	<1	$\gamma n \to p\pi^-\pi^0$, $n\pi^+\pi^-$	1500	30
ABHHM Collaboration	70	<6	$\gamma n \to p\pi^-\pi^0$, $n\pi^+\pi^-$	10000	31

The third type of experiments are bubble or streamer chamber experiments, in which the angles and momenta of all outgoing particles are measured over the full kinematic region. If the photon energy E_γ is unknown, one has 3 constraints from energy-momentum conservation so that distinction between reaction (1) and, say, reactions (7) and (8) is possible. Also reactions (4) and (6) have been investigated with this technique, using deuterium as a target. Here however, due to the neutral unmeasured particle in the final state there is no constraint from energy-momentum conservation if E_γ is unknown. Consequently, these reactions were not separated from the corresponding ones with additional neutral pions. Table 3 is a list of these experiments.

The total cross section for reaction (1)

$$\gamma p \to \pi^+ \pi^- p$$

from all experiments is shown in Fig. 4. From threshold it rises to a broad maximum at energies corresponding to c.m.s. energies E_{CM} of 1400–1600 MeV, and is falling again at $E_{CM} = 1700$ MeV. The rise

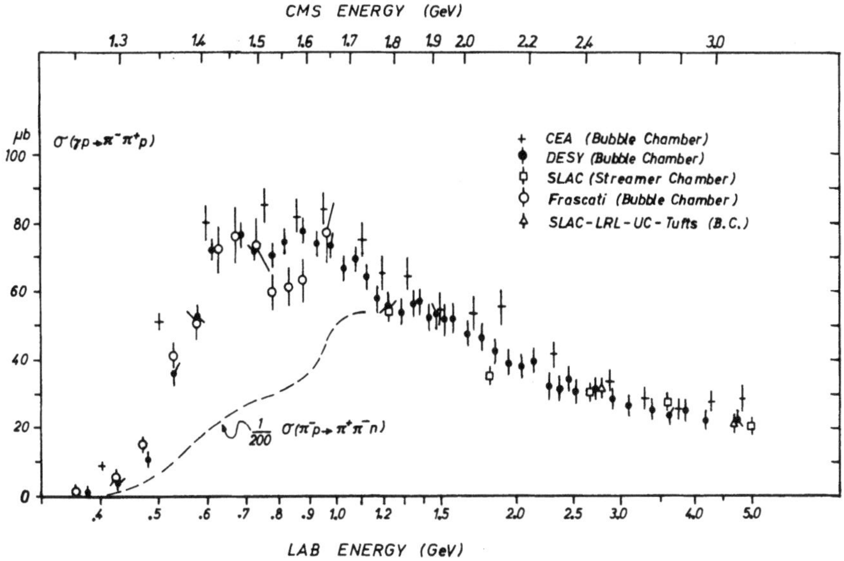

Fig. 4. Cross section of the reaction $\gamma p \to \pi^- \pi^+ p$. The curve shows 1/200 of $\sigma(\pi^- p \to \pi^- \pi^+ n)$ at the same c.m.s. energy

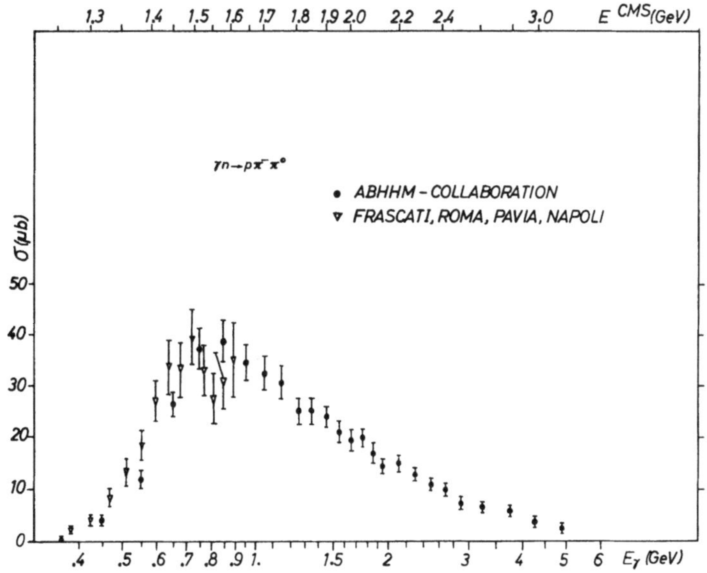

Fig. 5. Cross section of the reaction $\gamma n \to \pi^- \pi^0 p$, from measurements of $\gamma d \to \pi^- \pi^0 p p$. (Points above $E_\gamma \triangleq 1$ GeV are upper limits strictly because of possible additional undetected π^0's. The data of the ABHHM-Collaboration are preliminary.)

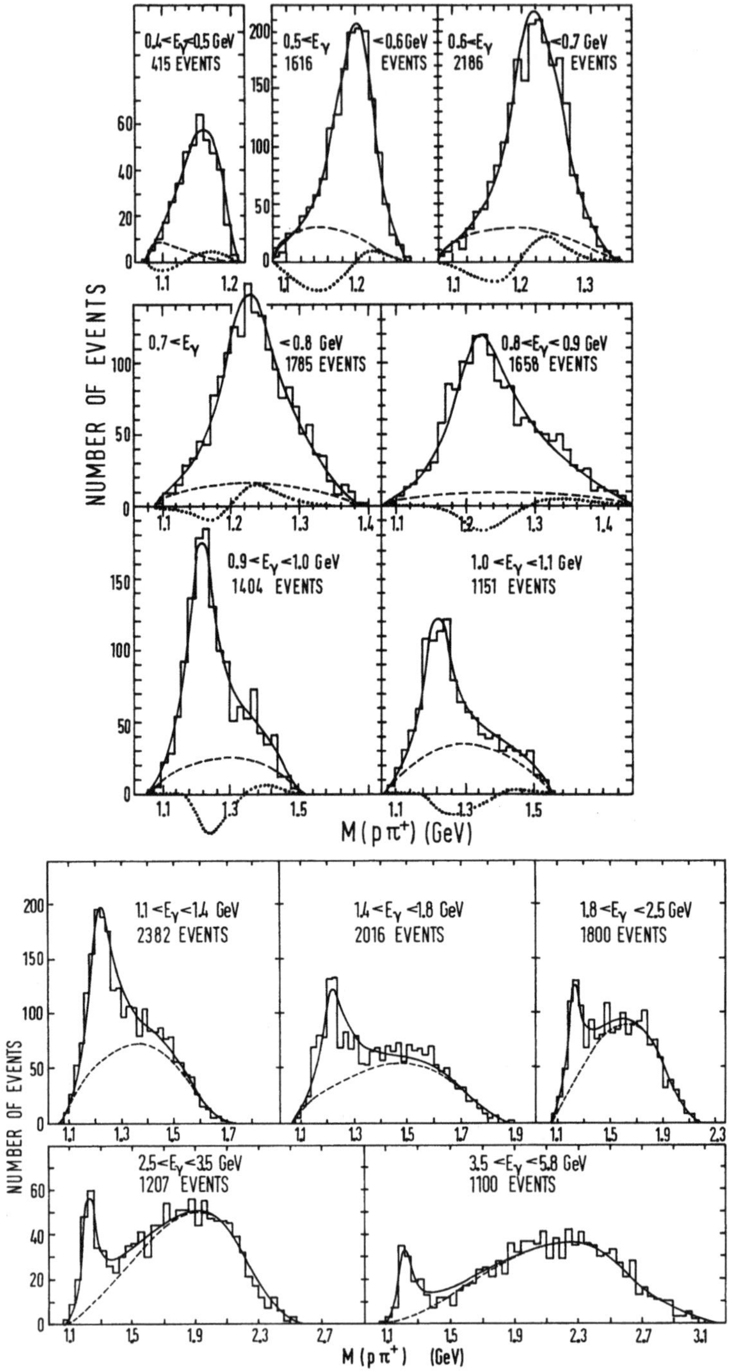

Fig. 6. Effective mass distribution of the $\pi^+ p$ system from reaction $\gamma p \to \pi^- \pi^+ p$ for various photon laboratory energies E_γ, from Ref. [10]. The curves are from fits as described in the text; the dashed and dotted curves show the amount of background and interference

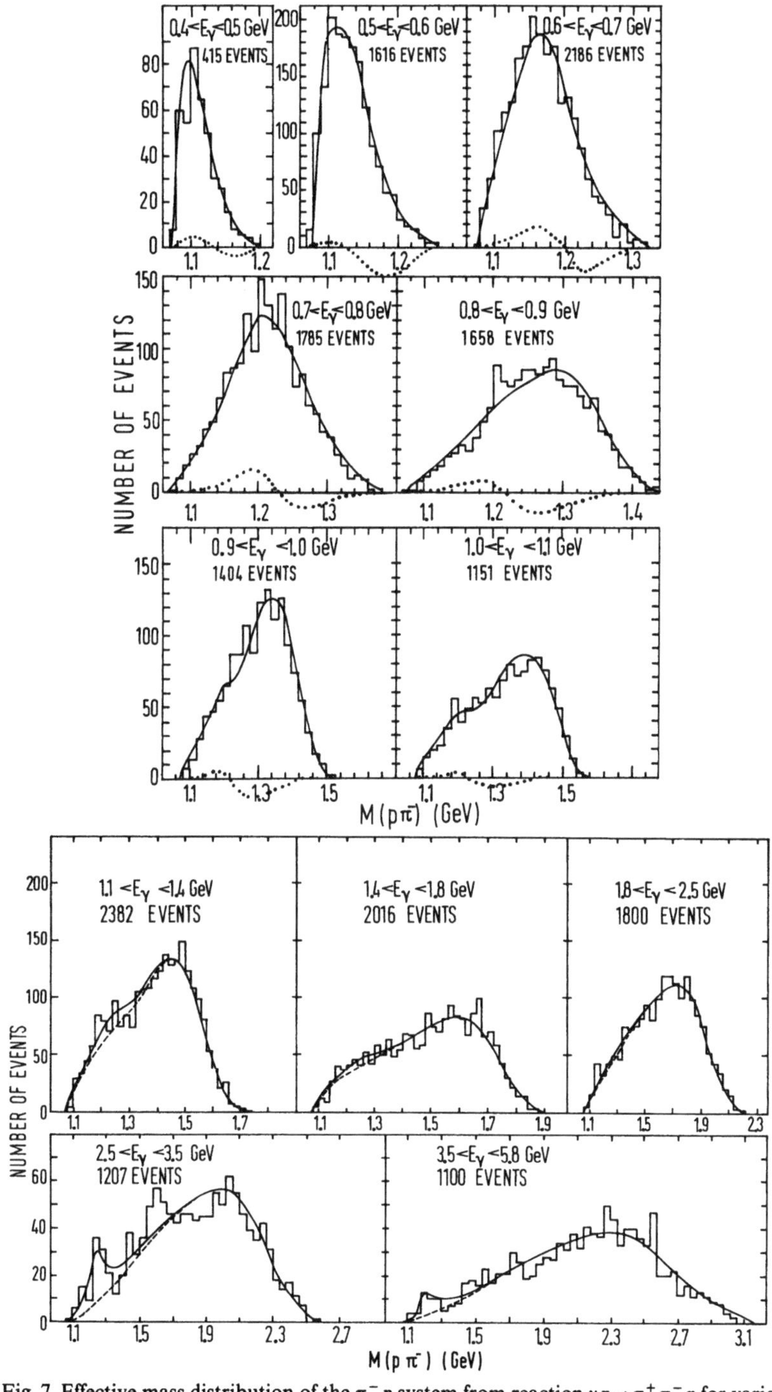

Fig. 7. Effective mass distribution of the $\pi^- p$ system from reaction $\gamma p \rightarrow \pi^+ \pi^- p$ for various photon laboratory energies E_γ, from Ref. [10]. The curves are from fits as described in the text; the dashed curves show the amount of background or interference

reflects just the $\pi\Delta(1236)$ threshold ($E_{CM} = 1372$ MeV) smeared by the width of the $\Delta(1236)$. This is compared in Fig. 4 with the reaction [32]

$$\pi^- p \to \pi^- \pi^+ n$$

whose final state is charge-symmetric to reaction (1). One sees that the behaviour of the two reactions as a function of E_{CM} is quite different. The pion-induced reaction does not have such a steeply rising cross section near $\pi\Delta$ threshold, and it continues to rise beyond the $E_{CM} \approx 1530$ MeV region.

Fig. 5 shows for comparison the cross section [30, 31] of reaction (6)

$$\gamma n \to \pi^- \pi^0 p .$$

This does not rise as high as the cross section for reaction (1). It peaks at about the same energy ($E_{CM} \approx 1500$ MeV) but falls down faster towards

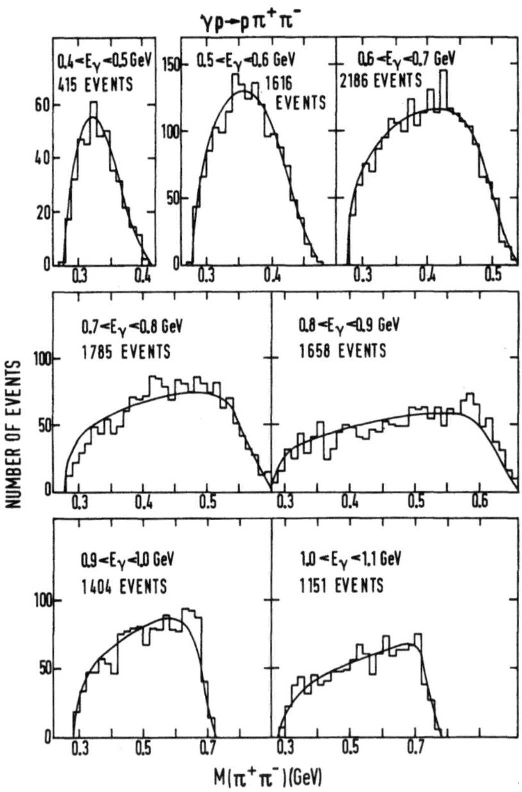

Fig. 8a

Fig. 8a and b. Effective mass distribution of the $\pi^+\pi^-$ system from reaction $\gamma p \to \pi^+\pi^- p$ for various photon laboratory energies E_γ, from Ref. [10]. The curves are from fits as described in the text; the dashed curves show the amount of background

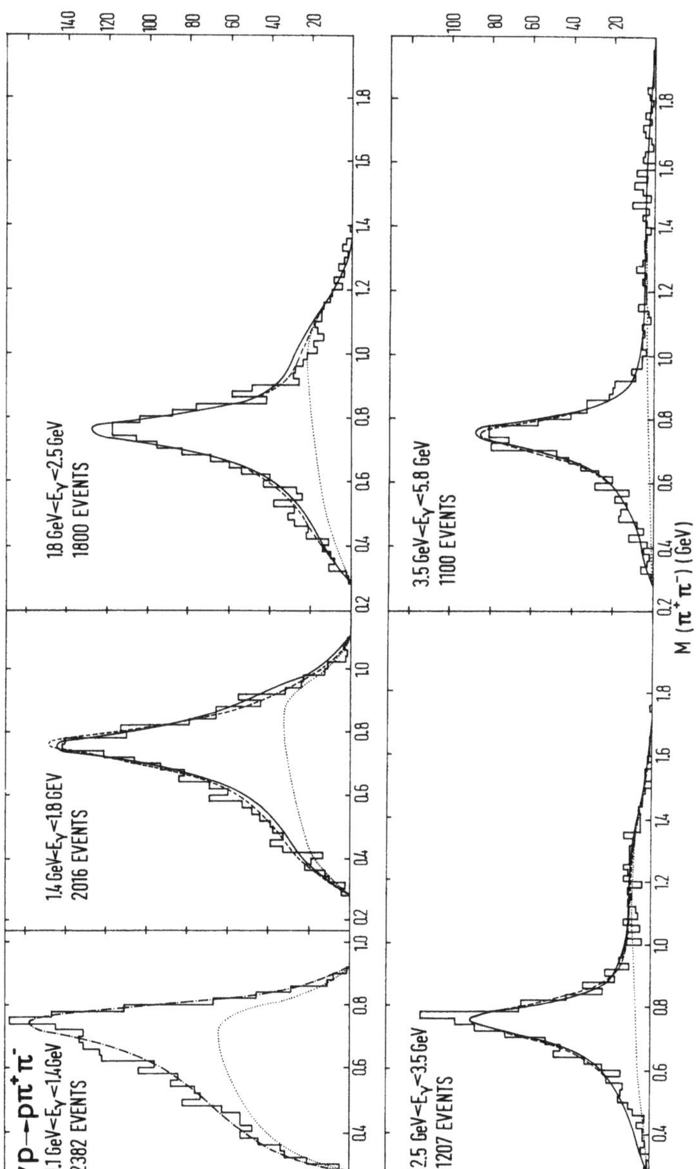

Fig. 8b

higher energies. Note that for $E_\gamma \gtrsim 1$ GeV the values given in Fig. 5 are upper limits only, because of the contributions from reactions with additional neutral pions.

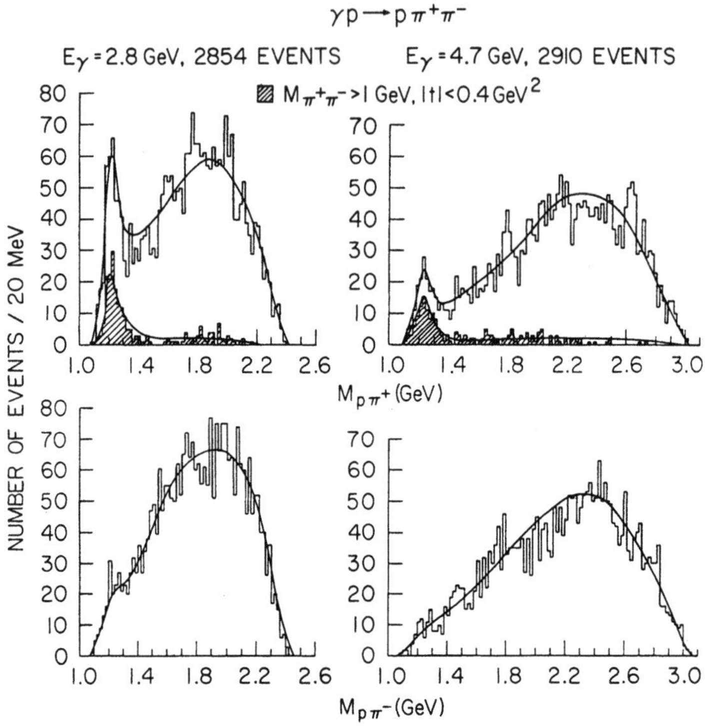

Fig. 9. Effective mass distribution of the $\pi^+ p$ and $\pi^- p$ systems from reaction $\gamma p \to \pi^+ \pi^- p$ at $E_\gamma = 2.8$ and 4.7 GeV, from Ref. [29]. The curves are from fits as described in the text.

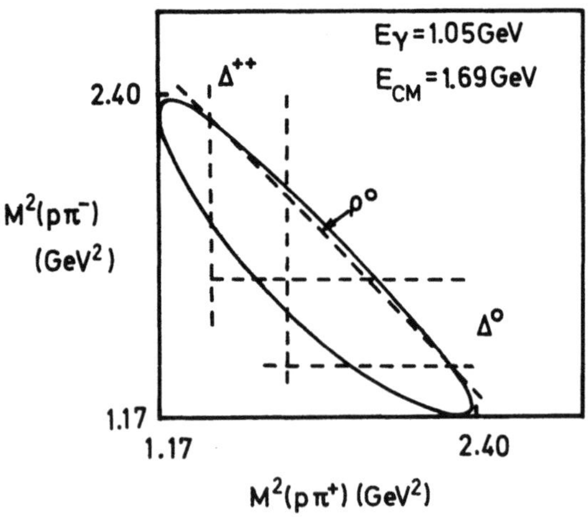

Fig. 10. Overlapping resonance bands on the Dalitz plot for the reaction $\gamma p \to \pi^+ \pi^- p$ at a photon laboratory energy of $E_\gamma = 1.05$ GeV, corresponding to a c.m.s. energy of $E_{CM} = 1.69$ GeV

The effective mass distributions of reaction (1) obtained by the ABBHHM Collaboration [10] are shown in Figs. 6, 7 and 8 for various regions of the photon energy E_γ. Fig. 9 shows mass distributions from the SLAC-Berkeley-Tufts experiment [29]. One observes strong Δ^{++} (1236) production at all energies. The $\pi^+ p$ and $\pi^- p$ mass distributions differ drastically, even at low energies where the phase space region available is not much larger than the width of the $\Delta(1236)$. The $\pi^+ p$ mass distribution always peaks near 1240 MeV, whereas one does not see any distinct indication of $\Delta^0(1236)\to\pi^- p$ production except at rather high energies. ϱ^0 production is seen in the $\pi^+\pi^-$ mass distribution at all energies above $E_\gamma = 1$ GeV. There is no evidence for any lowmass $\pi\pi$ enhancement.

Various fits have been made to these data, in order to find the contributions of

$$\Delta^{++}\to\pi^+ p$$

$$\Delta^0\to\pi^- p$$

$\pi^+\pi^-$ (*P* wave, in particular ϱ^0)

$\pi^+\pi^-$ (*S* wave)

phase space or peripheral (Drell-type [33]) background

in the $\pi^+\pi^- p$ final state. (We will frequently abbreviate $\Delta(1236)$ simply as Δ.) No indications for the production of other resonances have been observed in reaction (1). In the low-energy region discussed here, the various resonance bands overlap in the Dalitz plot. This is illustrated in Fig. 10. It is therefore important to take account of the reflections of the resonances, including the decay angular distributions of the Δ's and the $(\pi\pi)$ system; i.e. all of the resonance parameters including the spin density matrices have to be fitted simultaneously. Also there may be interference terms in the overlap regions. Further, since the Δ is a wide resonance, the mass dependence of the Δ width is important, in particular at high energy where the allowed phase space region is large.

Not all of these points have been (or could be) taken into account in the analysis of the various experiments. Therefore a compilation [2] of all published cross sections for reaction (9)

$$\gamma p\to\pi^-\Delta^{++} \tag{9}$$

looks somewhat confused (Fig. 11). A consistent determination [10, 34] of the various resonance cross sections at a range of photon energies shows a much clearer picture. Thus, the results in Fig. 12 have been obtained by allowing Δ^{++}, Δ^0 and phase space contributions below ϱ^0 production threshold ($E_\gamma = 1100$ MeV) and, in addition, a ϱ^0 contribution above threshold. The fits for $E_\gamma < 1100$ MeV were also done with *S* or *P*

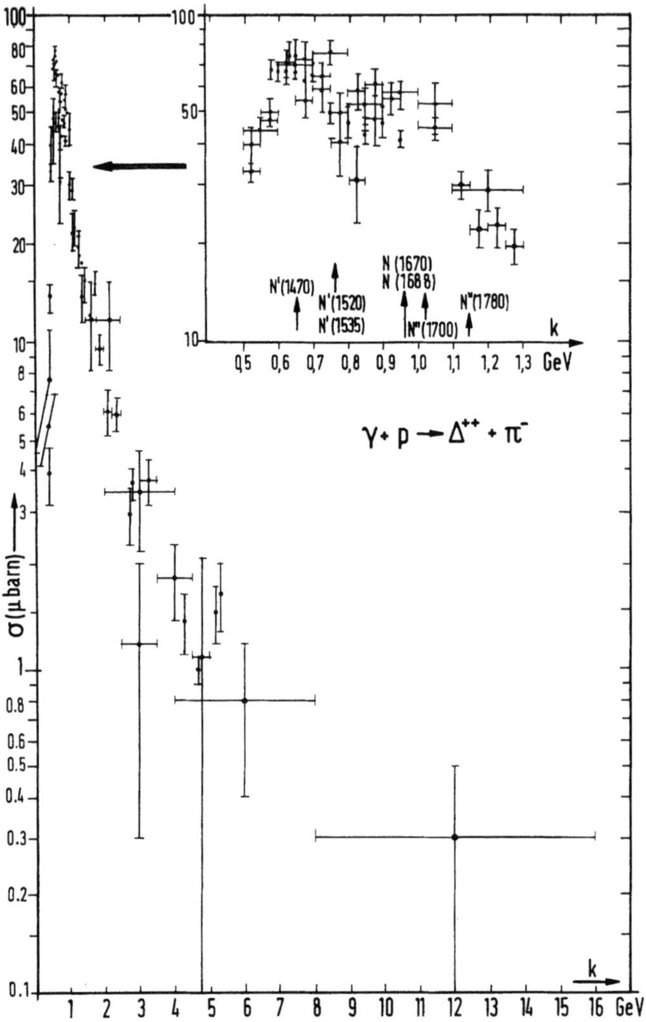

Fig. 11. Compilation of cross sections of the reaction $\gamma p \rightarrow \pi^- \Delta^{++}(1236)$ as a function of photon laboratory energy (k), from Ref. [2]

wave $\pi^+\pi^-$ states instead of a phase space contribution; this also resulted in good fits, with similar cross sections of the $\pi^+\pi^-$ S or P wave contribution as found for the phase space contribution. In other words, S or P wave $\pi^+\pi^-$ states, or a phase space contribution, could not be distinguished from each other. The cross section for Δ^{++} production was always found the same, independent of whether one simultaneously fitted a phase space, or a S or P wave $\pi^+\pi^-$ contribution. The curves

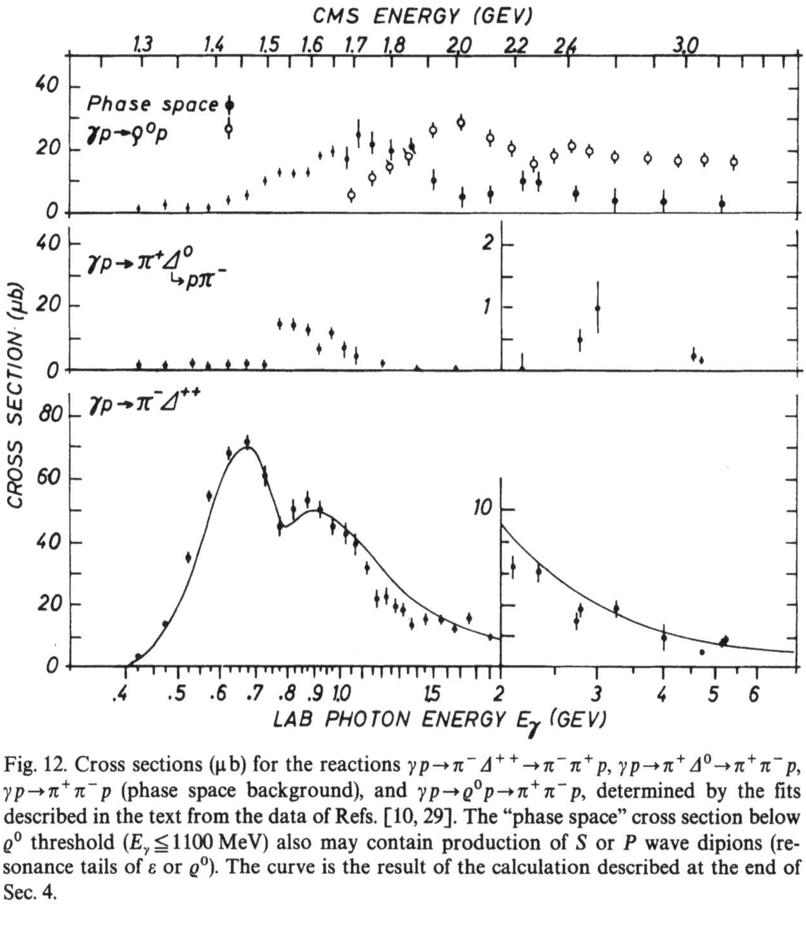

Fig. 12. Cross sections (μb) for the reactions $\gamma p \to \pi^- \Delta^{++} \to \pi^- \pi^+ p$, $\gamma p \to \pi^+ \Delta^0 \to \pi^+ \pi^- p$, $\gamma p \to \pi^+ \pi^- p$ (phase space background), and $\gamma p \to \varrho^0 p \to \pi^+ \pi^- p$, determined by the fits described in the text from the data of Refs. [10, 29]. The "phase space" cross section below ϱ^0 threshold ($E_\gamma \leqq 1100$ MeV) also may contain production of S or P wave dipions (resonance tails of ε or ϱ^0). The curve is the result of the calculation described at the end of Sec. 4.

in Figs. 6, 7 and 8 illustrate the results of the fit for the effective mass distributions.

As is seen from Fig. 12, the steep rise of the cross section $\sigma(\gamma p \to \pi^+ \pi^- p)$ with E_γ is nearly completely due to strong Δ^{++} production. There is < 10 μb of cross section for Δ^0 and/or phase space and for $\pi^+ \pi^-$ S-wave production below $E_\gamma = 700$ MeV ($E_{\mathrm{CM}} = 1500$ MeV). The ratio

$$\frac{\sigma(\gamma p \to \pi^- \Delta^{++} \to \pi^+ \pi^- p)}{\sigma(\gamma p \to \pi^+ \Delta^0 \to \pi^+ \pi^- p)} = \begin{cases} 9 & \text{for } I_s = 1/2, \text{ or } I_t = 1 \text{ with isovector photon} \\ 9/4 & \text{for } I_s = 3/2 \\ 1 & \text{for } I_t = 2 \end{cases}$$

(where I_s (I_t) is the $s(t)$ channel isospin) is found to be $\gtrsim 9$ experimentally for $E_\gamma \lesssim 700$ MeV, which indicates that $I = 1/2$ states in the s channel, or isovector t channel exchange, are dominant in the Δ production reaction

$$\gamma p \rightarrow \pi \Delta . \tag{10}$$

Δ production dominates up to $E_\gamma \approx 1200$ MeV. It has possible structure in the region of the second resonance bump in $\sigma_{tot}(\gamma p)$, at $E_\gamma = 700$–800 MeV corresponding to $E_{CM} = 1520$ MeV. At this energy, there seems to be a dip in the cross section for Δ^{++} production, associated with a bump in the Δ^0 cross section. This may however be spurious, since consideration of Figs. 6 and 7 reveals that at this energy the reflection of the Δ^{++} resonance in the $\pi^- p$ mass distribution has nearly the same form as the Δ^0 resonance itself, so that a distinction between Δ^{++} and Δ^0 production is then nearly impossible. The $\pi^+\pi^-$ P wave production cross section may reach a value of up to $\approx 20\,\mu$b in the $E_\gamma \approx 700$–800 MeV ($E_{CM} = 1520$ MeV) region; it is at most a few μb for higher energies up to the threshold ($E_\gamma = 1100$ MeV) for ϱ^0 meson production. From ϱ^0 threshold ($E_{CM} \mathrel{\hat{=}} 1700$ MeV) the cross section for the reaction

$$\gamma p \rightarrow \varrho^0 p \tag{11}$$

rises steeply to a maximum of about $25\,\mu$b in the region of $E_{CM} = 2000$ MeV. It then falls off slowly towards a rather energy-independent value of about $15\,\mu$b at high E_γ, assuming the features characteristic of elastic diffractive scattering (see Ref. 35). The Δ^{++} production cross section, on the other hand, is falling towards small values with increasing E_γ roughly like E_γ^{-2}.

4. The Reaction $\gamma p \rightarrow \pi^- \Delta^{++}(1236)$, Results and Interpretation

After having presented the behaviour of the cross section of the dominant quasi-two body low-energy photoproduction reaction

$$\gamma p \rightarrow \pi \Delta(1236) \tag{10}$$

as a function of E_γ in Fig. 12, we now turn to a more detailed discussion of this reaction. We already saw that this reaction predominantly has $I = 1/2$ in the s channel.

Naturally, one expects a strong influence of s channel resonances in the reaction $\gamma N \rightarrow \pi\pi N$ (see Fig. 13). In fact, various attempts to fit

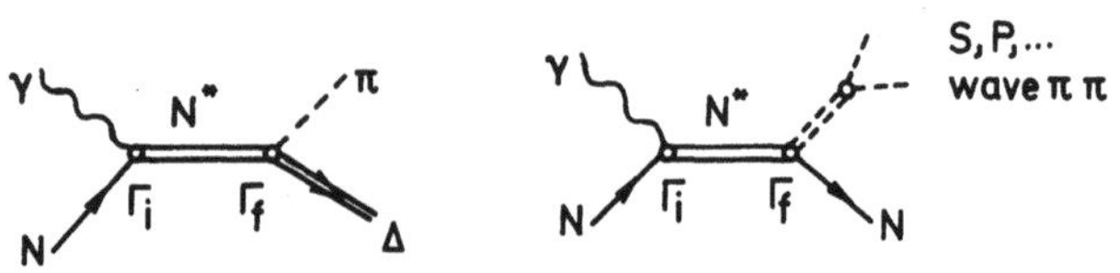

Fig. 13. s channel resonance excitation in the reactions $\gamma N \rightarrow \pi \Delta(1236)$ and $\gamma N \rightarrow \pi\pi N$

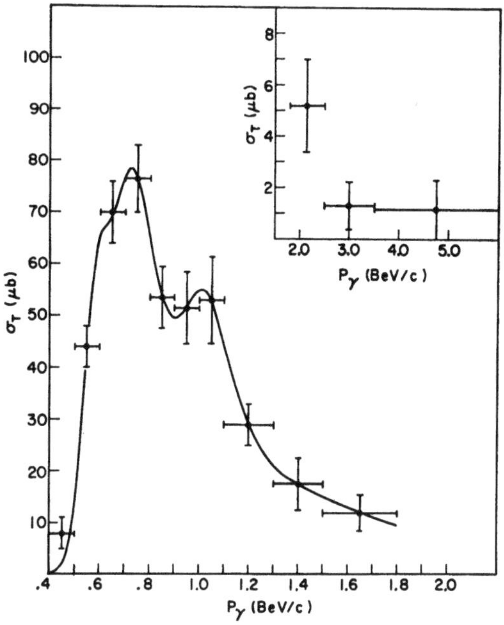

Fig. 14. Cross section for the reaction $\gamma p \to \pi^- \Delta^{++}(1236)$ as a function of photon laboratory energy (P_γ), fitted with an s channel resonance model with excitation of $N(1470)$, $N(1520)$, $N(1680)$ and $\Delta(1950)$. From Ref. [27].

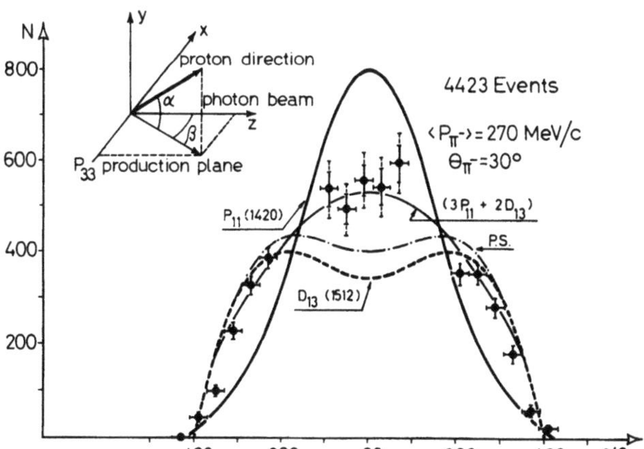

Fig. 15. Decay angular distribution of the Δ^{++} from reaction $\gamma p \to \pi^- \Delta^{++}$, relative to the normal on the production plane. The curves are calculated assuming excitation of $N(1470)$ ("P 11"), $N(1520)$ ("D 13"), or both of these resonances; "P.S." is the phase space distribution. The photon laboratory energy is between 550 and 850 MeV. From Ref. [21]

data of reaction

$$\gamma p \to \pi^- \Delta^{++} \tag{9}$$

with a pure "isobar model", i.e. s channel N^*'s in the intermediate state, were rather successful. Figs. 14 and 15 show two examples [21, 27].

We want, however, to first briefly review our present knowledge on $N^* \to \pi\pi N$ decays and on photoexcitation of N^*'s in order to see what estimates for the s channel resonance excitation cross sections in reaction (9) we can deduce from this.

Our knowledge of the inelastic decay modes of the non-strange baryon resonances comes from two principal sources. One is the elastic pion nucleon phase shift analysis [36] from which we know that all amplitudes with $I=1/2$, $J \leqq 5/2$ (except $J^P = \frac{3}{2}^+$) resonate below $E_{CM} \cong 1700$ MeV, with sizable inelasticities. The second source of information are the studies of the reaction

$$\pi N \to \pi\pi N$$

in this s channel resonance region. These are being pursued mainly in bubble chamber experiments with high statistics [36–39]. One analyses the reaction in partial waves and, assuming dominant two-body final-state interactions, also decomposes the 2-particle subsystems into partial waves. In this way one hopes to eventually determine the decay branching ratios of all N^*'s into channels like $\pi\Delta$, ϱN, or εN (we use "ε" and "ϱ" as a shorthand notation for a $\pi\pi$ system in the $I=J=0$ or $I=J=1$ state, respectively, but not necessarily with the mass of the ε or ϱ resonances). The preliminary information available today from these analyses is summarized in Table 4. In addition to the resonant amplitudes listed there, the 3-body partial wave analyses give evidence for background amplitudes, usually weaker but often still quite significant.

Next, we summarize in Table 5 the information available from single-pion photoproduction on the γN couplings of the N and Δ resonances [40, 41]. Here, one determines

$$\frac{\Gamma_{\gamma N}\Gamma_{\pi N}}{\Gamma^2} \propto \sum_{\text{helicities}} |T^{Jl}_{\text{res}}|^2$$

from the helicity amplitudes at resonance.

With these results one can now make very rough estimates of the contribution of the various N and Δ resonances in the s channel to the reactions

$$\begin{aligned} \gamma p &\to \pi^- \Delta^{++} \\ &\to \varrho^0 p \\ &\to \varepsilon p \text{ (with } \varepsilon \to \pi^+\pi^-) \end{aligned}$$

Table 4. *$\pi\pi N$ decays of N and Δ resonances*

Name	J^P	$l, 2I, 2J$ of πN decay	Γ(MeV)	fraction of $\pi\pi N$	$\pi\pi N$ channels (L)
$N(1470)$	$1/2^+$	P 11	200–400	40%	$\varepsilon N(S)$ probably dominant $\pi\Delta(P)/\varepsilon N(S)$ between 0 and 1 $\varrho N(P)$ possible
$N(1520)$	$3/2^-$	D 13	105–150	50%	$\pi\Delta(S)$ definite (prob. domin.) $\pi\Delta(D)$ probable $\varepsilon N(P)$ probable
$N(1535)$	$1/2^-$	S 11	50–160	mainly $N\eta$ only little $\pi\pi N$	$\varepsilon N(P)$ probable $\varrho N(S)$ seen recently, needs confirmation $\pi\Delta(D)$ zero or very small
$N(1670)$	$5/2^-$	D 15	105–175	60%	$\pi\Delta(D)$ ~40%, definite
$N(1688)$	$5/2^+$	F 15	105–180	40%	$\pi\Delta(P)$ ~25%, definite
$N(1700)$	$1/2^-$	S 11	100–400	not yet seen in $\pi\pi N$	
$N(1780)$	$1/2^+$	P 11	270–450		$\pi\Delta$ probable
$N(1860)$	$3/2^+$	P 13	310–450		ϱN possible
			nothing known yet about higher N's		
$\Delta(1236)$	$3/2^+$	P 33	120	no $\pi\pi N$	
$\Delta(1650)$	$1/2^-$	S 31	130–250	70%	$\pi\Delta(D)$ possible
$\Delta(1670)$	$3/2^-$	D 33	175–300		$\pi\Delta(S)$ definite
$\Delta(1690)$	$3/2^+$	P 33	270		$\pi\Delta(P)$ probable
$\Delta(1890)$	$5/2^+$	F 35	135–380		$\pi\Delta(F)$ possible
$\Delta(1910)$	$1/2^+$	P 31	230–420		$\pi N^{\star}_{1/2}(S)$ possible $\pi\Delta(P)$ possible
$\Delta(1950)$	$7/2^+$	F 37	140–220		$\pi\Delta(F) \simeq 50\%$ (definite) $\varrho\Delta$ probable [ϱN ~10% *]
$\Delta(2420)$	$11/2^+$	H 3,11	~310	>20%	[ϱN ~10% *]

* Rough estimate using vector dominance model (see Sec. 6).

using unitarity:

$$\sigma(E_{\mathrm{CM}}) \triangleq \frac{\pi}{k_{\mathrm{CM}}^2}(2J+1)\frac{m_{\mathrm{R}}^2 \Gamma_f(E_{\mathrm{CM}})\,\Gamma_{\gamma p}(E_{\mathrm{CM}})}{(m_{\mathrm{R}}^2 - E_{\mathrm{CM}}^2)^2 + m_{\mathrm{R}}^2 \Gamma^2(E_{\mathrm{CM}})}$$

where k_{CM} is the photon energy in the c.m.s., m_{R} the mass of the resonance and $\Gamma_f(E_{\mathrm{CM}})$ the partial width for the resonance to decay into the final

Table 5. γN couplings of N and Δ resonances, and contribution expected to $\gamma p \to \pi^+\pi^- p$.

Name	J^P	$l, 2I, 2J$ of πN decay	$\Gamma_{\gamma p}/\Gamma$ [%]	$\Gamma_{\gamma n}/\Gamma$ [%]	Max $\sigma(\gamma p \to\ )$ (μb) $\pi^-\Delta^{++}$	$\varrho^0 p$	εp
$N(1470)$	$1/2^+$	P 11	?? (guess $\lesssim 0.1$)		($\lesssim 1.5$)		($\lesssim 6$)
$N(1520)$	$3/2^-$	D 13	0.45	0.33	20		
$N(1535)$	$1/2^-$	S 11	~ 0.2	~ 0.4	0	~ 2	$\lesssim 4$
$N(1670)$	$5/2^-$	D 15	?? (guess $\lesssim 0.1$)		($\lesssim 5$)		
$N(1688)$	$5/2^+$	F 15	0.3	0 (?)	~ 10		
$\Delta(1236)$	$3/2^+$	P 33	0.52	0.52			
$\Delta(1950)$	$7/2^+$	F 37	~ 0.2	~ 0.2	~ 9	~ 3	

state considered (this contains Clebsch-Gordan coefficients projecting on the particular charge state observed). One then obtains the (gu)estimates given in the last three rows of Table 5.

The resonance contributions to the reaction

$$\gamma p \to \pi^- \Delta^{++}$$

estimated in this way are also compared in Fig. 16 with the observed cross section for this process. Although our estimate is a very rough one due to the large uncertainties in the partial widths $\Gamma_{\pi\Delta}$ and $\Gamma_{\gamma N}$ of the resonances, we can firmely conclude that there is no s channel resonance saturation in $\gamma p \to \pi^- \Delta^{++}$ at low energies. There must be very strong non-resonant background at low energies in this region.

In particular, the dramatic rise of $\sigma(\gamma p \to \pi^- \Delta^{++})$ between $E_\gamma = 500$ MeV and $E_\gamma = 700$ MeV ($E_{\rm CM} = 1350$–1480 MeV) certainly can not be *completely*

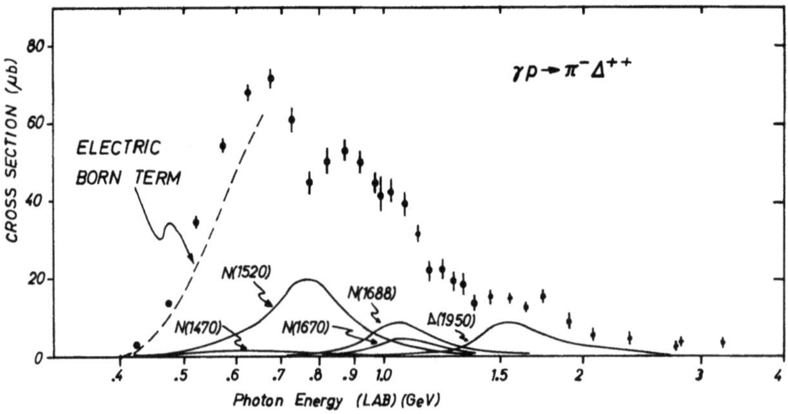

Fig. 16. s channel resonance contributions to $\sigma(\gamma p \to \pi^- \Delta^{++})$ expected from the rough estimates given in Table 5 compared with the experimental cross section for this reaction

explained by formation of $N(1470)$ with quantum numbers $J^P = 1/2^+$ (often called $P\,11$ resonance), for the following reasons:

(i) The observed cross section for the reaction $\gamma p \to$ "ε" $p \to \pi^- \pi^+ p$ (see Fig. 12 and associated discussion in the text) is too small to be compatible with the branching ratio [38]

$$\Gamma(N(1470) \to \pi \Delta)/\Gamma(N(1470) \to \varepsilon N) \lesssim 1 \quad \text{(see Table 4).}$$

(ii) A cross section of $\approx 60\,\mu$b for the reaction $\gamma p \to N^+(1470) \to \pi^- \Delta^{++}$ would imply, from isospin conservation and the elasticity of the $N(1470)$, a cross section of $\gtrsim 120\,\mu$b for $\gamma p \to N^+(1470) \to \pi^+ n$ (or, equivalently, a γN branching ratio of the $N(1470)$ of $\gtrsim 2\%$) which is excluded from the studies of single pion photoproduction.

(iii) The rise of $\sigma(\gamma p \to \pi^- \Delta^{++})$ occurs at too low energy and is too sudden, to be fitted reasonably with the $N(1470)$ resonance even if one assumes the mass to be only 1430 MeV.

What, then, is the nature of the large background amplitude near threshold? Fortunately there exists a time-honored model which turns out to describe rather well the behaviour of $\sigma(\gamma p \to \pi^- \Delta^{++})$ near threshold. In this model, the sudden rise of the cross section is due to a strong *contact interaction* leading to a predominantly S-wave $\pi^- \Delta^{++}$ final state, as shown in Fig. 17a. Originally proposed by *Cutkosky* and *Zachariasen* [42] and treated in the framework of the static theory [42, 43], this contact interaction was later also motivated [44, 45] from current algebra and PCAC, in the limit of a soft S-wave pion and from a phenomenological Lagrangian approach [46]. As had already been observed by *Stichel* and *Scholz* [47] the contact diagram is the leading graph near threshold in a gauge invariant electric Born term for the reaction

$$\gamma p \to \pi^- \Delta^{++}.$$

The corresponding graphs are shown in Fig. 17a–d in the order of decreasing importance at low energies. Pion exchange (Fig. 17b) is the leading t channel singularity; it leads to S, P and higher orbital angular

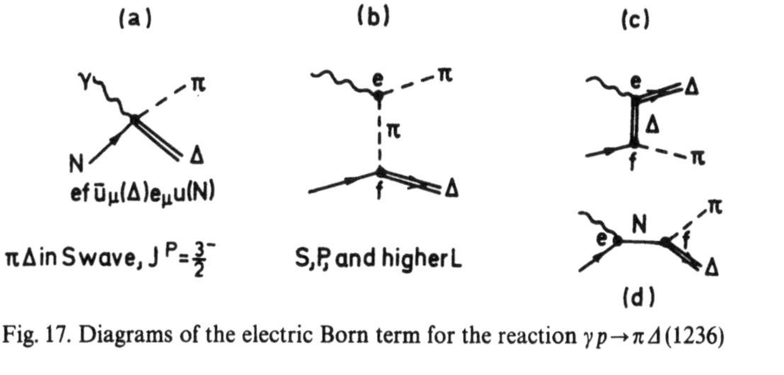

Fig. 17. Diagrams of the electric Born term for the reaction $\gamma p \to \pi \Delta(1236)$

momenta in the $\pi^- \Delta^{++}$ final state. The orbital current parts of Δ exchange in the u channel (Fig. 17c) and nucleon exchange in the s channel (Fig. 17d) formally serve to ensure gauge invariance of the total electric Born term [47] but are numerically unimportant. Thus in the Coulomb gauge in the c.m.s. the N exchange contribution vanishes and the Δ exchange contribution is very small at low energies as seen from Fig. 18. We therefore can ignore these parts of the amplitude. (At higher energies, the Δ exchange term does become large but when absorption corrections are applied, it again is negligible compared with the pion exchange term. See below for more details.)

The contributions of the electric Born diagrams to $\sigma(\gamma p \to \pi^- \Delta^{++})$ are shown in Fig. 18. Note that these contributions are not additive, due to interference terms.

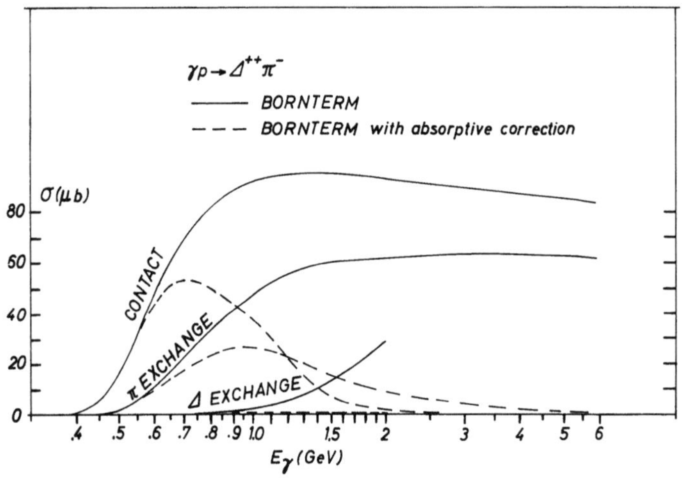

Fig. 18. Contribution of the electric Born diagrams (calculated in Coulomb gauge in the c.m.s.) to $\sigma(\gamma p \to \pi^- \Delta^{++})$.

It appears from this figure that the contact interaction amplitude can explain the rapid rise of $\sigma(\gamma p \to \pi^- \Delta^{++})$ from threshold, and that it even gives the correct magnitude of the cross section near threshold within about 20%. This then is apparently the source of the strong non-resonant background in $\sigma(\gamma p \to \pi^- \Delta^{++})$ near threshold. The smallness of $\sigma(\gamma p \to \pi^+ \Delta^0 \to \pi^+ \pi^- p)$ near threshold is expected in this picture since from the contact interaction alone this cross section is only 1/9 of the cross section $\sigma(\gamma p \to \pi^- \Delta^{++})$. The contact interaction leads to a nearly pure S wave $\pi \Delta$ state and therefore has $J^P = 3/2^-$. The production angular distribution $d\sigma/d\Omega$ of the final state has to be isotropic in the c.m.s. if *only* this state is present.

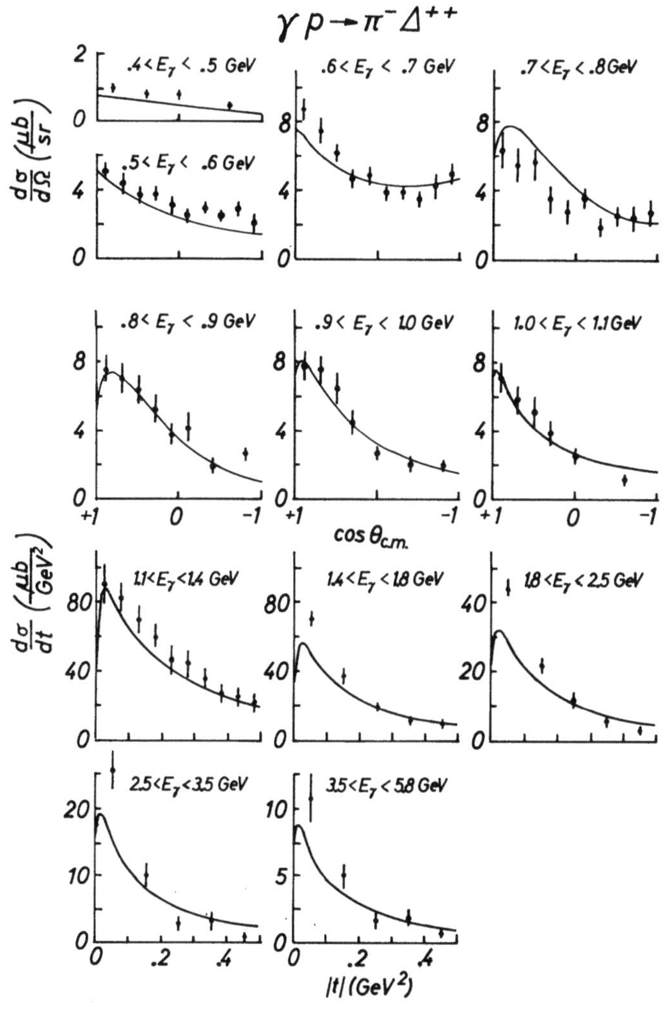

Fig. 19. C.m.s. production differential cross section $\mathrm{d}\sigma/\mathrm{d}\Omega$, and $\mathrm{d}\sigma/\mathrm{d}t$ (t = square of four-momentum transfer), of the reaction $\gamma p \to \pi^- \Delta^{++}$ (1236) obtained by fits as described in Sec. 3 from the data of Ref. [10]. The curves are the result of the calculation described at the end of Sec. 4

This production angular distribution is shown in Fig. 19. It is clearly indicative of *S and P* waves at small energies. This is even more clearly seen from Fig. 20 which shows the coefficients A_l of an expansion

$$\frac{\mathrm{d}\sigma}{\mathrm{d}\Omega}(\gamma p \to \pi^- \Delta^{++}) = \frac{1}{k_{\mathrm{CM}}^2} \sum_l A_l P_l(\cos\theta)$$

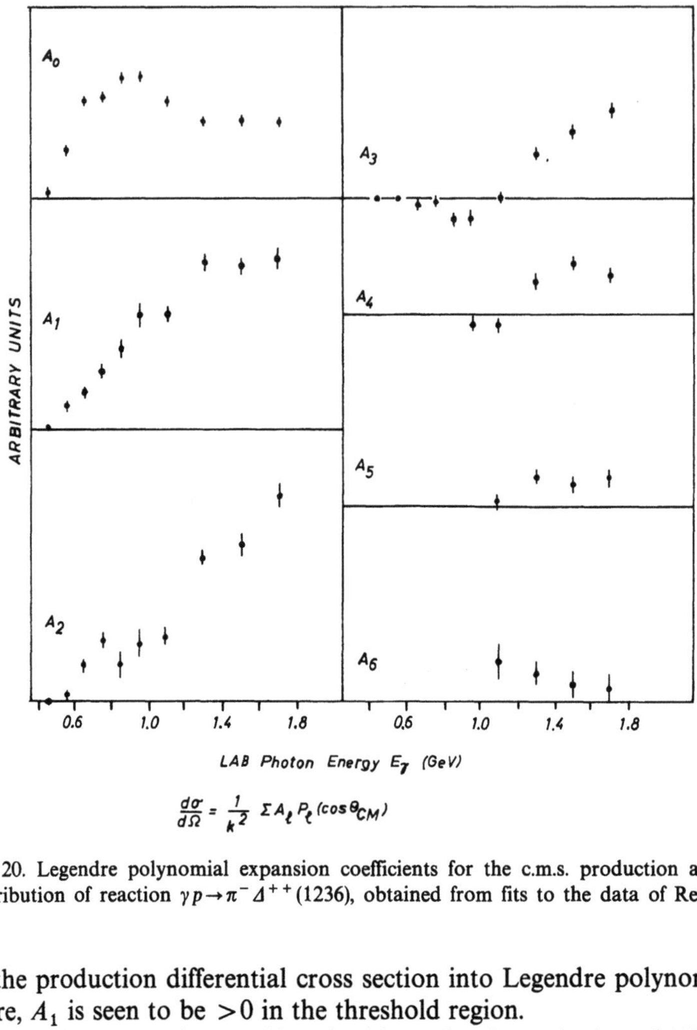

Fig. 20. Legendre polynomial expansion coefficients for the c.m.s. production angular distribution of reaction $\gamma p \to \pi^- \Delta^{++}(1236)$, obtained from fits to the data of Ref. [10]

of the production differential cross section into Legendre polynomials. Here, A_1 is seen to be >0 in the threshold region.

Can the P wave be considered evidence for the excitation of $N(1470)$ with $J^P = 1/2^+$ in the s channel? From [38]

$$\frac{\Gamma(N^+(1470)\to\pi^-\Delta^{++})}{\Gamma(N^+(1470)\to\text{“}\varepsilon\text{”}p)} = \frac{0.5\,\Gamma(N(1470)\to\pi\Delta)}{\Gamma(N(1470)\to\text{“}\varepsilon\text{”}N)} \leqq 0.5$$

and $\sigma(\gamma p \to \text{“}\varepsilon\text{”}p) < 10\,\mu\text{b}$ at the resonance energy of $E_\gamma = 680$ MeV (corresponding to $E_{CM} = 1470$ MeV) (see Fig. 12) we conclude that $\sigma(\gamma p \to N^+(1470) \to \pi^-\Delta^{++})$ must certainly be smaller than $5\,\mu\text{b}$ at resonance. Actually our rough guess in Table 5 even gave $\lesssim 1.5\,\mu\text{b}$. This is several times smaller than the real, non-resonant pion exchange contribution ($\sim 15\,\mu$b) expected at this energy (see Fig. 18). The latter also gives predominantly P waves. Although the calculation of the pion exchange at so small energies may not be considered quantitatively reliable, it is clear that in view of this background amplitude one would need data many times better than available at present, in order to draw a conclusion on the observation of the $N(1470)$ resonance in the reaction $\gamma p \to \pi^- \Delta^{++}$.

We can look back at the Legendre polynomial expansion coefficients of Fig. 20 to search for other structure perhaps indicative of resonance production. A_2 rises rather smoothly; and in fact we do not expect the $N(1520)$, $J^P = 3/2^-$ resonance to show up strongly there because its $\pi\Delta$ decay is predominantly S wave (see Table 4). A_3 is seen to change sign near $E_{\mathrm{CM}} = 1680$ MeV. This may be due to interference between the D wave $\pi\Delta$ decay of $N(1520)$ with $J^P = 3/2^-$, (or a D wave from pion exchange) and the $N(1688)$, $J^P = 5/2^+$ resonance which are both expected to contribute (see Table 5). To A_4, neither $N(1520)$ nor $N(1688)$ with its P wave $\pi\Delta$ decay contribute. It is actually seen to be small up to $E_\gamma \approx 1200$ MeV ($E_{\mathrm{CM}} \approx 1770$ MeV) where it begins to rise, perhaps indicating formation of $\Delta(1950)$ with $J^P = 7/2^+$.

From our rough estimate given in Table 5 we would expect a sizable contribution, of the order of $20\,\mu$b, from excitation of the $N(1520)$, $J^P = 3/2^-$ resonance. There is possible structure in the $E_{\mathrm{CM}} = 1520$ MeV region in the $\gamma p \to \pi^- \Delta^{++}$ data, namely the minimum in $\sigma(\gamma p \to \pi^- \Delta^{++})$ (see Fig. 12). The main inelastic decay of $N(1520)$ is $\pi\Delta$ S-wave. Thus it leads into the same state as does the contact interaction term. Therefore, the dip at $E_{\mathrm{CM}} = 1520$ MeV is suggestive of an interference between the dominant $J^P = 3/2^-$ background from the contact term, and excitation of the $N(1520)$ resonance.

Further evidence pointing to resonance structure in the $E_{\mathrm{CM}} = 1500$ bis 1600 MeV region comes from the polarization of the final state $\Delta^{++}(1236)$, as described by its helicity density matrix $\varrho^{\mathrm{H}}_{2\lambda 2\lambda'}$. Fig. 21 shows that $\varrho^{\mathrm{H}}_{33}$ (the probability that the Δ^{++} spin is aligned parallel to the c.m.s. momentum of the Δ^{++}) varies strongly in this region, indicating the presence of some rapidly varying amplitude there. $\varrho^{\mathrm{H}}_{33}$ also changes rapidly at $E_{\mathrm{CM}} \approx 1900$ MeV, and $\mathrm{Re}\,\varrho^{\mathrm{H}}_{31}$ has a broad, significant maximum centered at $E_{\mathrm{CM}} = 1950$ MeV, also probably indicating resonance structure.

We now describe an attempt to qualitatively interprete the $\gamma p \to \pi^- \Delta^{++}$ data at low energies, taking account of the Born terms and of s channel resonance contributions. This is an extension [34] of previous work [48, 49], with some modifications. The approach used is pictured in Fig. 22. It has the following features.

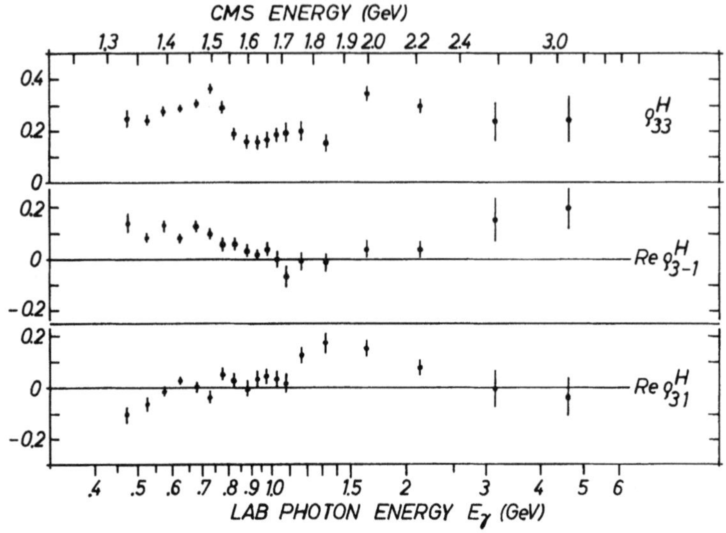

Fig. 21. Helicity density matrix elements of the Δ^{++} from the reaction $\gamma p \to \pi^- \Delta^{++}(1236)$ obtained from fits to the data of Ref. [10]

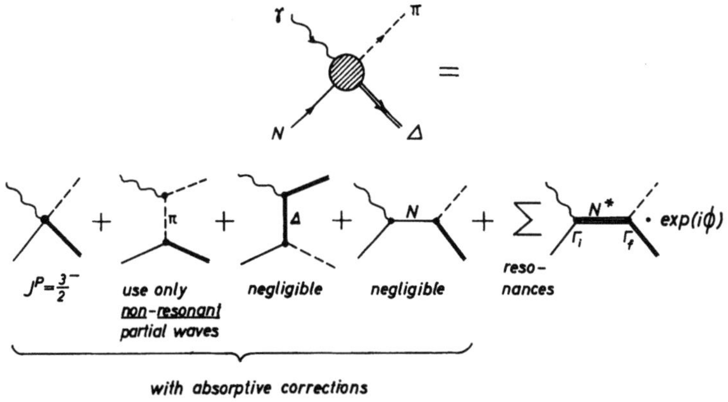

Fig. 22. A model to interprete the reaction $\gamma p \to \pi^- \Delta^{++}(1236)$

1. It contains the full contact term, with absorption corrections. The necessity to include the contact interaction, in addition to any $J^P = 3/2^-$ *resonance*, has already been discussed. Absorption corrections are needed since without them the amplitude grows too large above $E_\gamma \approx 700$ MeV (see Fig. 18).

2. The pion exchange term is analyzed in partial waves. The non-resonant partial waves are added into the amplitude. Absorption corrections are applied here also.

3. The resonant partial waves are parametrized in the energy region near resonance by Breit-Wigner amplitudes with an additional free, constant phase factor $\exp(i\Phi)$ for each resonance [48, 50]. At higher energies outside the resonance, the pion exchange partial wave amplitudes are taken instead.

Absorption effects in the final $\pi\Delta$ state may be expected even at threshold energies, but there is no reliable way to calculate them at low energy. For the absorption factor in the helicity amplitudes, to extrapolate the absorption effects to low energy, we therefore simply use the asymptotic formula [51, 52]

$$S_{ff}^{1/2} = \{1 - C_\Delta \exp[-(J-1/2)^2/2A_\Delta q_\Delta^2]\}^{\frac{1}{2}},$$

with q_Δ the c.m.s. momentum of a $\pi\Delta$ state at given total energy, $C_\Delta = 1$, and $A_\Delta = 8\ \text{GeV}^{-2}$. Above ϱ^0 threshold we also add initial state absorption, in the spirit of the vector dominance model [53]. This we describe by

$$S_{ii}^{1/2} = \{1 - C_\varrho \exp[-(J-1/2)^2/2A_\varrho q_\varrho^2]\}^{\frac{1}{2}},$$

with $C_\varrho = 1$ and $A_\varrho = 5.5\ \text{GeV}^{-2}$ (see Ref. [35]). The values of the absorption parameters C and A are chosen such as to imply a total $\pi^-\Delta^{++}$ cross section of 38 mb as expected from the additive quark model, and a $\varrho^0 p$ cross section of 27 mb as deduced from high energy ϱ^0 photoproduction data [35].

The partial wave amplitudes arising from the pion exchange diagram are shown, with and without absorption corrections, in Fig. 23.

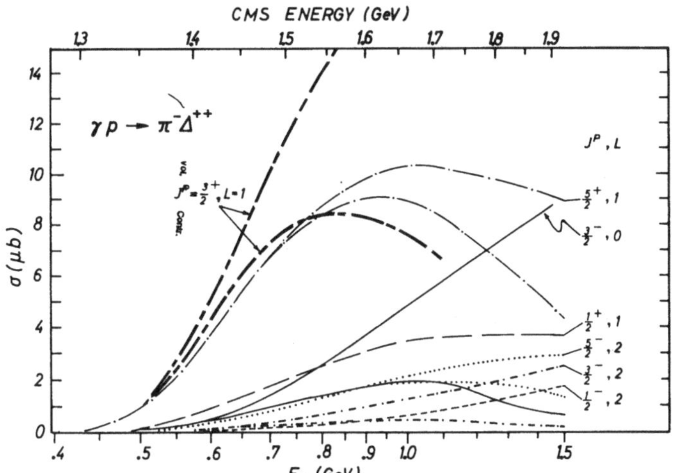

Fig. 23. Contributions of the partial waves from the pion exchange diagram to the cross section of the reaction $\gamma p \to \pi^- \Delta^{++}(1236)$. J^P is spin and parity, and L the final $\pi\Delta$ orbital angular momentum. The splitting of the curves shows the effect of the absorption corrections.

Table 6. *Resonance parameters used in the model calculation*

Name	M (MeV)	Γ (MeV)	J^P	λ_i = total initial helicity	L of $\pi\Delta$	Φ (deg)	$\sigma(\mu b)$ at resonance
$N(1470)$	1435	200	$1/2^+$	1/2	1	0	1.5
$N(1520)$	1520	115	$3/2^-$	3/2	0	240	9

Only the two lowest s channel resonances have been included so far, with parameters as given in Table 6. A detailed fit of these parameters to the data has not yet been tried.

The result of the calculation described above for the total cross section $\sigma(\gamma p \to \pi^- \Delta^{++})$ is shown by the curve in Fig. 12. It is in good

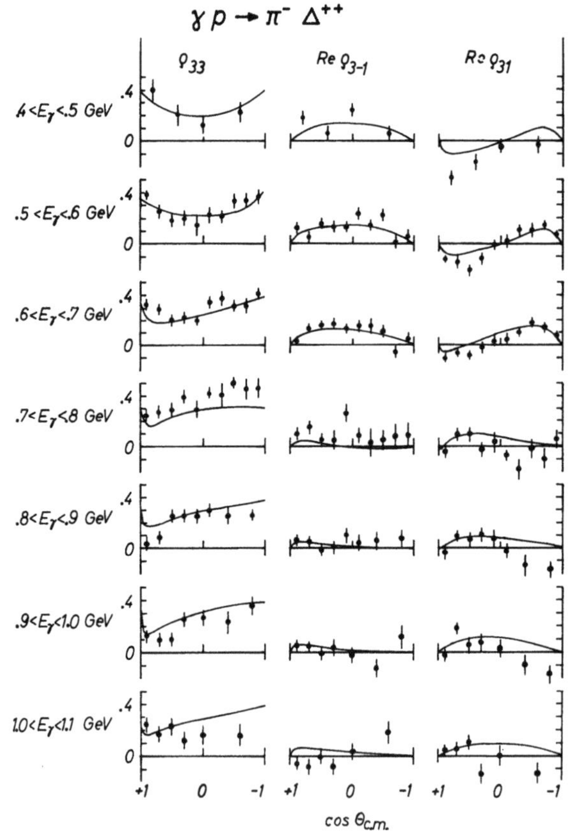

Fig. 24. Helicity density matrix elements of the Δ^{++} from the reaction $\gamma p \to \pi^- \Delta^{++}(1236)$ as a function of the cosine of the c.m.s. production angle, determined by fits to the data of Ref. [10]. The curves are the result of the calculation described at the end of Sec. 4

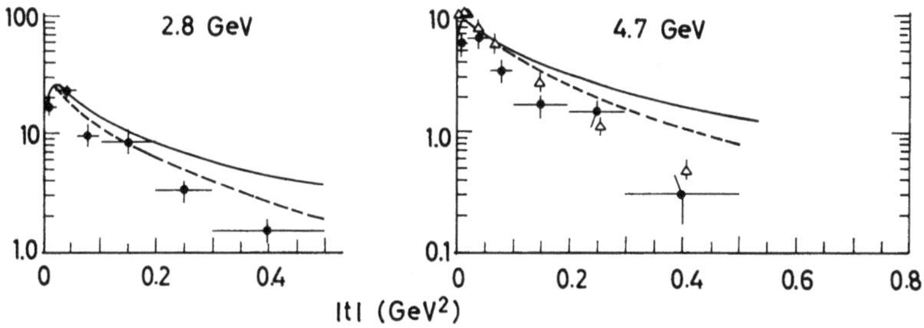

Fig. 25. Differential cross section $d\sigma/dt(\mu b/GeV^2)$ of the reaction $\gamma p \to \pi^- \Delta^{++}(1236)$ as a function of four-momentum transfer squared $|t|$ (GeV^2) at photon laboratory energies of 2.8 and 4.7 GeV. The curves are predictions from the Born term with absorption corrections with $C=0.8$ (———) and $C=1$ (- - - -). From Ref. [29]

agreement with the magnitude and the energy dependence of this cross section. Also the general features of the differential production cross section and of the helicity density matrix $\varrho^{H}_{2\lambda 2\lambda'}(\cos\theta)$ as a function of the c.m.s. production angle θ are reasonably well reproduced with this simple model calculation, as shown by the curves in Figs. 19 and 24.

Finally it is satisfying to observe that this model has the correct high energy behaviour [29]. This is given by the Born terms alone; with absorption corrections actually only the pion exchange term remains non-negligible. The degree of agreement is illustrated in Figs. 19 and 25. Rho Regge pole exchange may also be present, but the residuum functions are unknown.

5. Further Discussion of the Reaction $\gamma N \to \pi \Delta(1236)$

In the preceding section, the reaction $\gamma p \to \pi^- \Delta^{++}$ was interpreted to have a strong, real, non-resonant $J^P = 3/2^-$ partial wave near threshold from the contact interaction, in addition to pion exchange and s channel resonance contributions. An interesting test would be afforded by a measurement of the reaction

$$\gamma n \to \pi^0 \Delta^0 \to \pi^0 \pi^- p. \qquad (12)$$

Here, pion exchange as well as the electric contact interaction term is absent. s channel resonance contributions are left. We expect a much smaller cross section than for the reaction $\gamma p \to \pi^- \Delta^{++}$. The preliminary results [31] available indeed indicate this to be the case.

Important additional information could be supplied by experiments with polarized photons [54]. For example, with unpolarized photons there is little hope to disentangle a possible $N(1470)$ resonance contribution from the Born term P wave contribution (see also Section 4). The Born P waves, coming mainly from the pion exchange diagram, vanish if the reaction plane is perpendicular to the polarization plane of the photon, due to the $\boldsymbol{e} \cdot \boldsymbol{q}_\pi$ electric dipole term in the amplitude, where $\boldsymbol{e}$ is the polarization vector of the photon, and $\boldsymbol{q}_\pi$ the c.m.s. momentum of the pion. So if $J^P = 1/2^+$ P waves are still present in this kinematic situation, they must have a different origin, presumably resonant.

A measurement of the reaction $\gamma p \to \pi^- \Delta^{++}$ in the $E_\gamma = 570$–800 MeV region ($E_{CM} = 1400$–1540 MeV) has in fact been done with linearly polarized photons at *Stanford* [19] (see also Table 1), but without sufficient precision to decide on the $N(1470)$ excitation. The result of an azimuthal asymmetry of $\lesssim 10\%$ however suggests [19, 48] contributions from $N(1470)$ and/or $N(1520)$. A study of the reaction $\gamma N \to \pi \Delta$ with polarized photons was also done, at much higher energies $E_\gamma = 2.8$ GeV and $E_\gamma = 4.7$ GeV, by the SLAC-Berkeley-Tufts Collaboration [29]. They found agreement with the pure Born term model (with absorptive corrections) at small four-momentum transfer $|t| \lesssim 0.5$ GeV^2 (see Fig. 25).

We now present a summary of our conclusions on the reaction $\gamma N \to \pi \Delta$ in the low-energy region.

1. s channel resonance excitation alone does not explain this reaction.

2. Near threshold the reaction is dominated by a strong contact interaction amplitude in the $J^P = 3/2^-$ state.

3. There are indications, in the cross section and the helicity density matrix, of rapidly varying amplitudes in the regions of $E_{CM} \approx 1520$ MeV, $E_{CM} \approx 1680$ MeV and probably $E_{CM} \approx 1950$ MeV.

4. A quantitative interpretation has to wait until better data are available, in particular from experiments with polarized photons. It is important to measure in the whole kinematical region of the $\pi\pi N$ final state.

6. Vector Meson Production

Above ϱ^0 threshold at $E_\gamma = 1100$ MeV ($E_{CM} \approx 1700$ MeV), with increasing photon energy E_γ, ϱ^0 production soon becomes the dominant process contributing to the overall reaction $\gamma p \to \pi^+ \pi^- p$. In this energy region, the main background to ϱ^0 production seems to be a Drell-type amplitude [33], i.e. pion exchange with πN diffraction scattering at the lower vertex [25, 55, 56]. The steep rise of the cross section of the ϱ^0 production reaction

$$\gamma p \to \varrho^0 p \tag{11}$$

from threshold reminds one of S wave production, but the production angular distributions are found to be anisotropic. We seem to have a superposition of a

diffractive amplitude + s channel resonances.
(non-resonant)

The diffractive contribution to the cross section is expected to rise smoothly, with a threshold behaviour yet unknown, towards a constant value. The resonances are interfering with this smooth diffractive background. No double counting presumably occurs, since according to the Freund-Harari conjecture [57, 58] the diffractive part is connected via duality with a non-resonant background while the resonances are connected with Regge pole exchange terms.

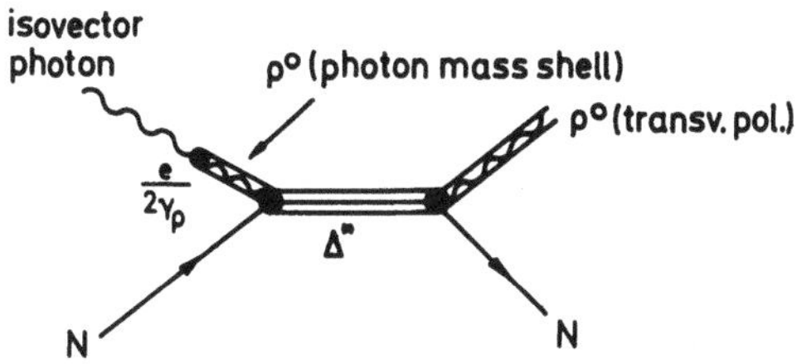

Fig. 26. Excitation of Δ resonances ($I=3/2$) in the reaction $\gamma N \to \varrho^0 N$, in the vector dominance model.

In order to estimate the contribution of the resonances to reaction (11) one may use the vector dominance model [59, 60]. The γN coupling of Δ resonances, with isospin $I=3/2$, proceeds in this model via $\varrho^0 N$ coupling and the known direct $\gamma\varrho^0$ coupling (Fig. 26). One therefore expects to see in reaction (11) *all* Δ resonances which are seen in *any* photoexcitation process. From the γN couplings one can roughly infer the $\varrho^0 N$ couplings, correcting for the kinematical mass dependence of phase space and centrifugal barrier factors when one goes from the photon to the ϱ^0 mass shell. Resonance contributions estimated in this way are given in Table 7.

The measurements of the ϱ^0 production cross section as a function of photon energy are not yet precise enough to observe resonance effects

Table 7. *Estimated contribution* [59, 60] *of photoproduced* Δ *resonances to* $\sigma(\gamma p \to \varrho^0 p)$

Name	J^P	$\sigma(\gamma p \to \varrho^0 p)$ at resonance
$\Delta(1950)$	$7/2^+$	$\sim 3\ \mu b$
$\Delta(2420)$	$11/2^+$	$\sim 0.6\ \mu b$
$\Delta(2850)$		(indication seen [61] in $\gamma p \to p\pi^0$)

of the order of 1 μb as predicted. Only indications of resonance structure (e.g. Fig. 12 of Ref. [10]) have been seen.

It may be expected that ϱ^0 production by linearly *polarized* photons is more sensitive to resonance effects. In diffraction the polarization vector does not change and accordingly the decay plane of the ϱ^0 is expected to be predominantly parallel ($\sigma_{\parallel}$) to the polarization plane of the photon. In resonant ϱ^0 photoproduction, on the other hand, the polarization vector may change. *Criegee et al.* [26] have studied the decay azimuthal anisotropy

$$\Sigma = \frac{\sigma_{\parallel} - \sigma_{\perp}}{\sigma_{\parallel} + \sigma_{\perp}}$$

of the ϱ^0 in the c.m.s. energy range $E_{CM} = 1.9$–2.7 GeV. Similar measurements were made by the SLAC-Berkeley-Tufts Collaboration [62] and by *Diambrini-Palazzi et al.* [63]. As one sees from Fig. 27, the observed anisotropies Σ are nearly equal to 1 as excepted for pure diffraction. This does, however, not exclude resonance contributions, since unfortunately, as shown by *Schildknecht* and *Schrempp* [60], this experiment is insensitive to resonances that are excited and decay by *magnetic* transitions. From the quark model [60, 64] one actually expects the Δ resonances of Table 7 to be magnetically excited, and the vector dominance hypothesis then also predicts magnetic decay into $\varrho^0 N$.

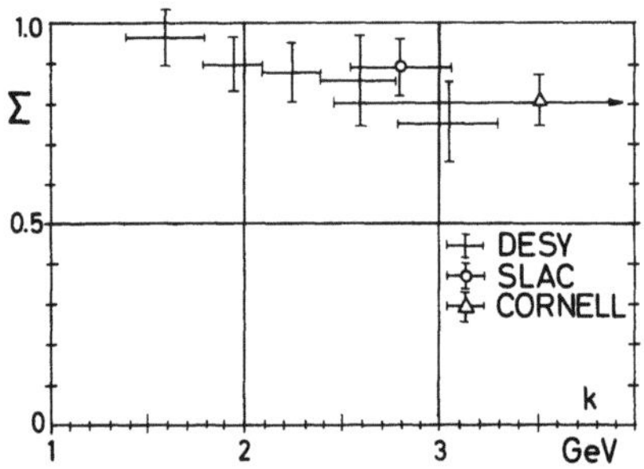

Fig. 27. Polarization anisotropy $\Sigma = (\sigma_{\parallel} - \sigma_{\perp})/(\sigma_{\parallel} + \sigma_{\perp})$ in the reaction $\gamma p \to \varrho^0 p$ with four-momentum transfer squared $|t| < 0.4\ \mathrm{GeV}^2$, as a function of photon laboratory energy (k). From Refs. [26, 62, 63]

To study ϱ photoproduction in the *absence* of a diffractive background, one may turn to the charge-exchange reaction

$$\gamma n \to \varrho^- p \to \pi^- \pi^0 p. \tag{13}$$

Fig. 28 shows effective mass distributions of the $\pi^-\pi^0$ system from the reaction $\gamma d \to \pi^- \pi^0 pp$, studied by the ABHHM Collaboration [65].

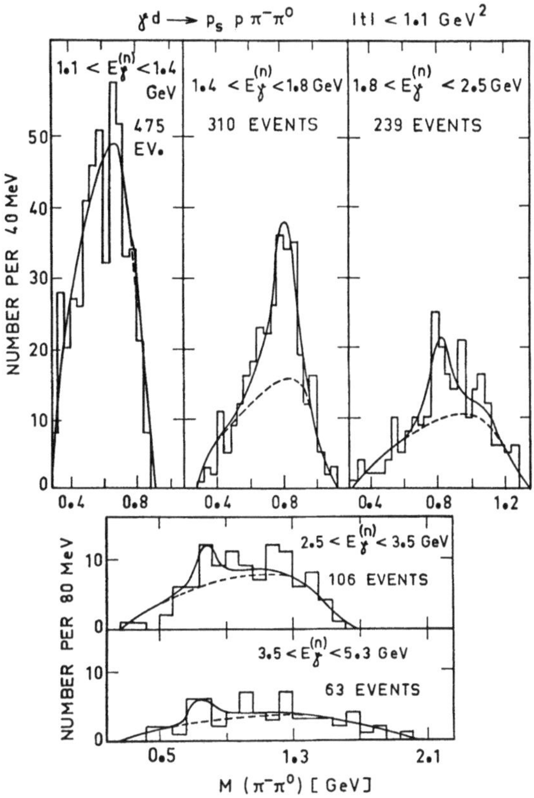

Fig. 28. Effective mass distribution of the $\pi^-\pi^0$ system from reaction $\gamma d \to \pi^-\pi^0 pp$, for various photon energies in the rest frame of the reacting neutron. The curves are from fits showing ϱ^- production and background. From Ref. [65]

ϱ^- production is observed, although it is not nearly as dominant as in the reaction $\gamma p \to \varrho^0 p \to \pi^+\pi^- p$ (Fig. 8). The ϱ^- photoproduction cross section of reaction (13) deduced from this experiment is shown and compared with ϱ^0 photoproduction in Fig. 29. It appears that ϱ^- production has a much smaller cross section, even at low energies in the s channel resonance region, but the errors are very large due to deuteron effects. For pure s channel isospin states we would have

$$\frac{\sigma(\gamma n \to \varrho^- p)}{\sigma(\gamma p \to \varrho^0 p)} = \begin{cases} 2 & \text{for } I_s = 1/2, \text{ pure isovector (or pure isoscalar) current} \\ 0.5 & \text{for } I_s = 3/2 \text{ (isovector current).} \end{cases}$$

The results of Fig. 29 support the notion [58] that diffraction is the dominant process in ϱ^0 photoproduction at energies $E_\gamma \gtrsim 2$ GeV. Better data are needed, however, for a more detailed study.

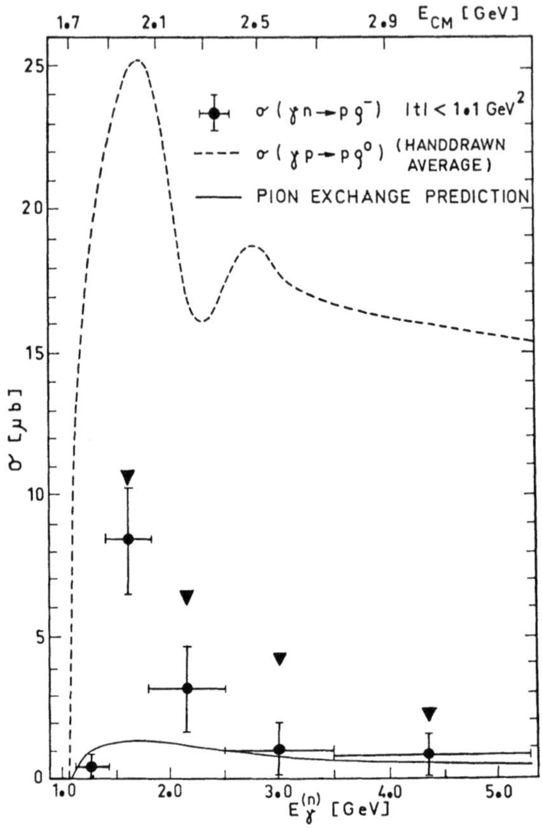

Fig. 29. Cross section of the reaction $\gamma n \to \varrho^- p$ as a function of photon energy in the n rest frame. The experimental points are for $|t| < 1.1\ \mathrm{GeV}^2$, the triangles show upper limits for all $|t|$. The dashed curve shows $\sigma(\gamma p \to \varrho^0 p)$ for comparison, the full curve is the pion exchange prediction. From Ref. [65]

We now turn to another process, namely ω production in the reaction

$$\gamma p \to \omega p \to \pi^+ \pi^- \pi^0 p. \tag{14}$$

This has at low energies been observed in bubble [10, 66, 67] and streamer [28] chamber experiments, but only about 1300 events of reaction (14) were obtained so far below $E_\gamma = 3$ GeV. The background under the ω is easily subtracted thanks to the small width of 12 MeV of the ω. The resulting cross section for ω production as a function of photon energy E_γ is shown in Fig. 30. It rises steeply from threshold ($E_\gamma = 1100$ MeV), keeps a value of about 7 μb between 1200 and 2200 MeV, and then falls off gradually to a value of about 3 μb at high energies. As for ϱ^0 photoproduction one expects contributions from both diffrac-

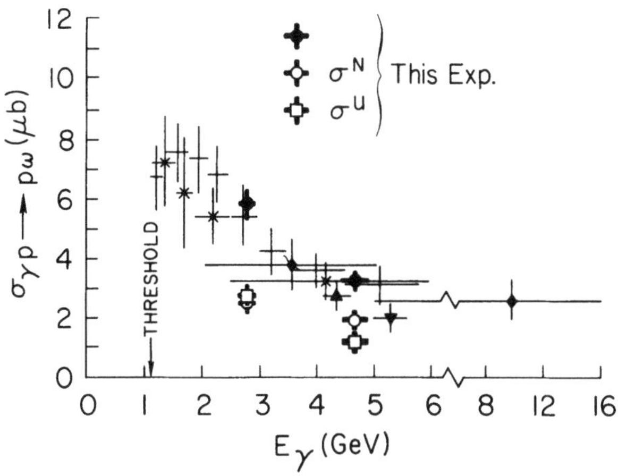

Fig. 30. Compilation of cross sections of the reaction $\gamma p \to \omega p$ as a function of photon laboratory energy E_γ. $\sigma^N(\sigma^U)$ are the contributions from natural (unnatural) J^P exchange in the t channel (for $|t| < 1$ GeV^2). From Ref. [67]

tion (= non-resonant background) and s channel resonances. According to the vector dominance model, in this reaction one should observe all the N resonances (with $I = 1/2$) that are excited by isoscalar photons. The data are however much too imprecise to resolve any possible structure in the cross section below 3 GeV.

The maximum of the ω production cross section in the $E_\gamma = 1200$–2200 MeV region actually agrees within the wide experimental errors with what is expected from pion exchange, assuming that there is in addition a diffractive part to the cross section. From the data [67] obtained with polarized photons at 2.8 and 4.7 GeV it is possible to separate the natural and unnatural spin-parity exchange contributions (see Fig. 30). The magnitude of the unnatural exchange contribution is then found to agree with the magnitude of the pion exchange term.

The differential cross section for ω production changes from approximate isotropy in $E_\gamma = 1100 - 1400$ MeV (see Fig. 19 of Ref. [10]) to a distribution sharply peaked in the forward direction at energies above 2 GeV. The forward peak may be mainly due to diffraction and pion exchange. The backward differential cross section, of the order of 0.1 μb/ster in the 2–3 GeV region, is somewhat larger than expected from these exchanges alone.

We end with an observation [68] of an apparent peculiarity in the total γp cross section (extrapolated from electroproduction measurements [3]) when compared with the "average" pion-proton cross section

$$\sigma_{tot}(\pi p) = (1/2)[\sigma_{tot}(\pi^+ p) + \sigma_{tot}(\pi^- p)]$$

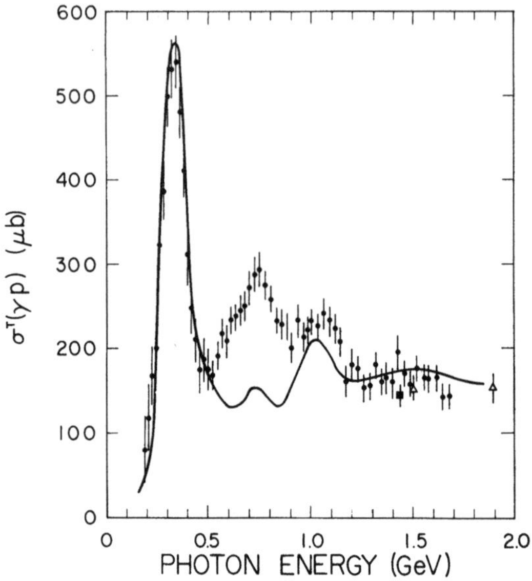

Fig. 31. Total photoproduction cross section $\sigma_{tot}(\gamma p)$ (μb), extrapolated from electroproduction data; from Ref. [3]. The curve is 1/200 of the mean $\pi^+ p$ and $\pi^- p$ total cross sections (Ref. [68])

(Fig. 31). It appears remarkable that the ratio $\sigma_{tot}(\gamma p)/\sigma_{tot}(\pi p)$ is approximately the same from threshold up to $E_\gamma = 20$ GeV. The only exception is a region near $E_\gamma = 750$ MeV ($E_{CM} = 1510$ MeV) where $\sigma_{tot}(\gamma p)$ has a relatively larger peak than $\sigma_{tot}(\pi p)$. The latter may indicate the presence of some possibly interesting effect, coupled more strongly to photons (and in the vector dominance picture, to the rho meson) than to pions.

References

1. *Fischer, H.*: Experimental Photoproduction at Intermediate Energies; invited talk at this symposium.
2. Adapted from: *Joos, P.*: Compilation of Photoproduction Data above 1.2 GeV; DESY-HERA 70–1 (Sept. 1970), unpublished.
3. *Bloom, E. D., et al.*: SLAC-PUB-653 (1969), unpublished.
4. *Caldwell, D. O., et al.*: Phys. Rev. Letters **25**, 609 (1970).
5. *Meyer, H., et al.*: Phys. Letters **33** B, 189 (1970).
6. Aachen-Berlin-Bonn-Hamburg-Heidelberg-München Collaboration. Phys. Letters **27** B, 474 (1968).
7. *Ballam, J., et al.*: Phys. Rev. Letters **21**, 1544 (1968).
8. *Ballam, J., et al.*: Phys. Rev. Letters **23**, 498 (1969); Phys. Rev. Letters **21**, 1541 (1968); and private Communication from *Wolf, G.*
9. Cambridge Bubble Chamber Group: Phys. Rev. **155**, 1477 (1967).

10. Aachen-Berlin-Bonn-Hamburg-Heidelberg-München Collaboration. Phys. Rev. **175**, 1669 (1968); see also *Meyer, A.F.A.*: Nucl. Phys. B **9**, 244 (1969).
11. Aachen-Berlin-Bonn-Hamburg-Heidelberg-München Collaboration. Phys. Rev. **188**, 2060 (1969).
12. *Costa, S., et al.*: Nuovo Cimento **45** A, 696 (1966).
13. *Eisenberg, Y., et al.*: Phys. Rev. Letters **22**, 669 (1969).
14. *Gialanella, G., et al.*: Nuovo Cimento **63** A, 892 (1969).
15. *Beneventano, M., et al.*: Nuovo Cimento **38**, 1054 (1965).
16. *Allaby, J.V., Lynch, H.L., Ritson, D.M.*: Phys. Rev. **142**, 887 (1966).
17. *Hauser, M.G.*: Phys. Rev. **160**, 1215 (1967).
18. *Burg, J.P., et al.*: Nuovo Cimento **56** A, 1142 (1968).
19. *Morrison, R.W., Drickey, D.J., Mozley, R.F.*: Nuovo Cimento **56** A, 409 (1968).
20. *Del Fabro, R., et al.*: Phys. Rev. **139** B, 701 (1965).
21. *Diambrini-Palazzi, G., et al.*: Phys. Rev. Letters **21**, 1463 (1968).
22. *Ukai, K., et al.*: INSJ-107 (Tokyo Univ.) (1967), unpublished.
23. *Shanks, W.L.*: Thesis, Calif. Inst. of Technology (1968), unpublished.
24. *Franklin, A.D., et al.*: Phys. Rev. Letters **13**, 491 (1964).
25. *Eisenhandler, E., et al.*: Phys. Rev. Letters **20**, 758 (1968).
26. *Criegee, L., et al.*: Phys. Rev. Letters **25**, 1306 (1970).
27. *Cambridge Bubble Chamber Group*: Phys. Rev. **163**, 1510 (1967).
28. *Davier, M., et al.*: Phys. Rev. D **1**, 790 (1970).
29. *Bingham, H.H., et al.*: Phys. Rev. Letters **25**, 1223 (1970).
30. *Piazza, A., et al.*: Lett. Nuovo Cimento **3**, 403 (1970).
31. Aachen-Bonn-Hamburg-Heidelberg-München Collaboration, private communication.
32. See, for example: *Rosenfeld, A.H., Söding, P.*: N^* Studies via $\pi^- p$ Inelastic Reactions; in: Pion Nucleon Scattering, p. 47. Ed. *G.L. Shaw* and *D.Y. Wong*. New York 1969.
33. *Drell, S.D.*: Phys. Rev. Letters **5**, 278 (1960); Rev. Mod. Phys. **33**, 458 (1961).
34. For more details, see: *Lüke, D.*: Thesis, University of Hamburg (1971), unpublished.
35. *Wolf, G.*: Photoproduction of Vector Mesons; invited talk at this symposium.
36. For references and discussion, see: Particle Data Group, Review of Particle Properties. Phys. Letters **33** B, 1 (1970); and *Plano, R.J.*: $S=0$ Baryon Resonances; in: Proceedings of the Lund International Conference on Elementary Particles, Lund 1969, p. 321.
37. *de Beer, M., et al.*: Nucl. Phys. B **12**, 599 (1969).
38. *Saxon, D.H., Mulvey, J.H., Chinowsky, W.*: Phys. Rev. D **2**, 1790 (1970).
39. *Brody, A.D., et al.*: Phys. Rev. D **3**, 2619 (1971).
40. *Walker, R.L.*: Single Pion Photoproduction in the Resonance Region; in: Proceedings of the International Symposium on Electron and Photon Interactions at High Energies, p. 23. Liverpool: Phys. Rev. **182**, 1729 (1969).
41. *Pfeil, W., Schwela, D.*: Springer Tracts in Mod. Phys. **55**, 213 (1970).
42. *Cutkosky, R.E., Zachariasen, F.*: Phys. Rev. **103**, 1108 (1956).
43. *Carruthers, P., Wong, H.*: Phys. Rev. **128**, 2382 (1962).
44. *Ebata, T.*: Phys. Rev. **154**, 1341 (1967); *Carruthers, P., Huang, H.W.*: Phys. Letters **24** B, 464 (1967); *Adler, S.L., Weisberger, W.I.*: Phys. Rev. **169**, 1392 (1968); *Kamal, A.N.*: Phys. Rev. **171**, 1556 (1968); *Bhargava, S.C.*: Phys. Rev. **174**, 1969 (1968); *Narayanaswamy, P., Renner, B.*: Nuovo Cimento **53** A, 107 (1968).
45. *Dosch, H.G., Gordon, D.*: Ann. Phys. **50**, 472 (1968).
46. *Huang, H.W.*: Phys. Rev. **174**, 1799 (1968).
47. *Stichel, P., Scholz, M.*: Nuovo Cimento **34**, 1381 (1964); (see also *Ciocchetti, G., et al.*: Lett. Nuovo Cimento **1**, 21 (1969)).
48. *Lüke, D., Scheunert, M., Stichel, P.*: Nuovo Cimento **58** A, 234 (1968).
49. See e.g.: *Harari, H.*: Comparison of Experiments with Models for Photoproduction, in: Proceedings of the 1967 International Symposium on Electron and Photon Interactions at High Energies, p. 337. Stanford 1967.

50. *McVoy, K. W.:* Nuclear Resonance reactions and S-matrix analyticity, in: Fundamentals in Nuclear Theory, p. 419. Lectures, Trieste 1966.
51. *Gottfried, K., Jackson, J. D.:* Nuovo Cimento **34**, 735 (1964).
52. *Locher, M. P., Sandhas, W.:* Z. Physik **195**, 461 (1966).
53. *Kramer, G., Schilling, K., Stodolsky, L.:* Nucl. Phys. B **5**, 317 (1968).
54. *Beder, D. S., Söding, P.:* UCRL-17867 (1967), unpublished; – *Böckmann, K., et al.:* Z. Physik **202**, 477 (1967).
55. *Söding, P.:* Phys. Letters **19**, 702 (1966); – *Krass, A. S.:* Phys. Rev. **159**, 1496 (1967).
56. *Bingham, H. H., et al.:* Phys. Rev. Letters **24**, 955 (1970).
57. *Freund, P. G. O.:* Phys. Rev. Letters **20**, 235 (1968).
58. *Harari, H.:* Phys. Rev. Letters **20**, 1395 (1968); – *Harari, H., Zarmi, Y.:* Phys. Rev. **187**, 2230 (1969).
59. *Joos, H., Schildknecht, D.:* Nuovo Cimento **57** A, 593 (1968).
60. *Schildknecht, D., Schrempp, B.:* DESY 70/8 (1970).
61. *Buschhorn, G., et al.:* Phys. Rev. Letters **20**, 230 (1968).
62. *Ballam, J., et al.:* Phys. Rev. Letters **24**, 960 (1970); **24**, 1467 (1970).
63. *Diambrini-Palazzi, G., et al.:* Phys. Rev. Letters **25**, 478 (1970).
64. *Renard, F. M.:* Nucl. Phys. B **2**, 537 (1967).
65. Aachen-Bonn-Hamburg-Heidelberg-München Collaboration. Nucl. Phys. B **21**, 93 (1970).
66. *Cambridge Bubble Chamber Group.* Phys. Rev. **155**, 1468 (1967).
67. *Ballam, J., et al.:* Phys. Rev. Letters **24**, 1364 (1970).
68. *Leith, D. W. G. S.:* Private communication.

Dr. *Dieter Lüke* and
Dr. *Paul Söding*
II. Institut für Experimentalphysik der Universität and DESY
2000 Hamburg, Germany

Photoproduction of Vector Mesons

G. WOLF

Contents

I. Introduction

One of the first processes studied [1–3] when CEA (1963) and DESY (1964) came into operation was the photoproduction of ϱ^0 mesons,

$$\gamma p \to p \varrho^0 . \tag{1}$$

The production features which were observed showed striking similarities to those of, e.g., elastic πN scattering; the momentum transfer distribution falls off exponentially with approximately the same slope; the total cross section for ϱ^0 production above 2 GeV is almost constant with energy.

This behaviour came not as a complete surprise to everyone. *Berman* and *Drell* [4], for example, by noting that the isovector part of the photon and the ϱ^0 have the same quantum numbers, had predicted that the production process will be mainly diffractive.

As we shall see all experimental data presently available – in particular the results obtained with linearly polarized photon beams – are consistent with the idea that reaction (1) proceeds predominantly through diffraction scattering. Accepting the diffraction hypothesis, reaction (1) becomes an extremely useful tool to study e.g. the spin dependence of diffraction scattering. For this purpose reaction (1) is much better suited than, e.g., elastic πN scattering because the decay angular distribution of the ϱ^0 which is readily accessible to experiment tells us the alignment of the ϱ^0 spin. In order to obtain the equivalent information for elastic πN scattering requires a polarized target and a measurement of the recoil polarization – a rather complicated experiment*.

Besides the ϱ^0 there are the vector mesons ω and φ which can be photoproduced without quantum number exchange via

$$\gamma p \to p\omega \tag{2}$$

and

$$\gamma p \to p\varphi \tag{3}$$

Thus these reactions can also be used to study the diffraction mechanism.

Another interesting aspect of the photoproduction of vector mesons has to do with the vector meson dominance model (VDM) [6] which links these processes to the total hadronic photon nucleon cross section and to vector meson production in e^+e^- annihilation.

In the following we shall discuss the experimental data on reactions (1–3) and compare them with various VDM predictions.

II. Rho Photoproduction, $\gamma p \to p\varrho^0$

1. Experimental Techniques

We start with ϱ^0 photoproduction which has been studied extensively. In spite of the large amount of data which exists on this reaction there is still a controversy on the size of the ϱ^0 production cross section both for experimental and theoretical reasons. In order to understand the experimental problems it is necessary to have a look at the various experimental setups which were used for these measurements.

Table 1 lists the more recent rho production experiments together with some relevant parameters.

a) Trackchamber Experiments

Several experiments have used a trackchamber device such as a hydrogen bubble chamber, or a streamer chamber with a hydrogen target. In these experiments all three particles of the final state $p\pi^+\pi^-$ are detected and momenta and angles measured. Since the beam direction is known one has three constraints. The contamination by events from other reactions is found to be negligible. The smallest momentum transfer for which reaction (1) can be analyzed with reliability using a trackchamber is

* As *Höhler* and *Strauss* have shown recently [5] there exists also the possibility of using phase shift results together with dispersion-theory in order to learn about the spin dependence of elastic diffractive πN scattering at high energies.

Table 1. *More recent experiments measuring cross sections for* $\gamma p \rightarrow p \varrho^0$

	technique	beam	photon energy (GeV)	particles detected	number of rho events above 2 GeV	remarks
CEA collab. [1]	HBC	bremsstrahlung	$\leqq 6$		500	final state $p\pi^+\pi^-$ is 3 (or 4) – constrained; background from other reactions is negligible
DESY [2] (ABBHHM)			$\leqq 6$	p, π^+, π^-	2000	
SLAC [7]		quasi monochromatic	5–8		150	
Weizmann [8]			4.3		300	
SLAC-Berkeley-Tufts [9]		Laser beam, lin. pol., approx. monochromatic	2.8; 4.7		4000	
SLAC [10]	Streamerch.	bremsstrahlung	2–16		400	
DESY [11] (*Blechschmidt et al.*)	counter set up	tagged γ's	3–5	π^+, π^-	350	1 constraint, $\sin^2\theta$ assumed
Cornell [12]		bremsstrahlung	4–8			0 constraints; inelastic background cannot be excluded; $\sin^2\theta$ assumed
DESY-MIT [13]			3–7		10^5	
SLAC (*Bulos et al.* [14])		monochromatic	9		1400	1 constraint, missing mass technique; resolution not good enough to separate ϱ and ω production
SLAC (*Anderson et al.* [15])		bremsstrahlung	6–18	p and missing mass	many	

$|t| = 0.02\,\text{GeV}^2$ (Ref. [9]). In general, however, losses occur up to $|t| \leqq 0.05\,\text{GeV}^2$.

The largest number of rho mesons with such a technique was detected by the SLAC-Berkeley-Tufts Collaboration [9]. In this experiment the photon beam was produced by Compton backscattering of an intense laser beam on high energy electrons (see Fig. 1). This is a two-body process; consequently for a given laser beam, k_i, and incident electron energy, E_e, the energy, E_γ, of the scattered photon beam depends only on the scattering angle θ. Collimation of the scattered beam then allows to monochromatize the beam. Also, if the incident laser beam is polarized, either linearly or circularly, the photons backscattered near $\theta = 0°$ will be polarized in the same fashion.

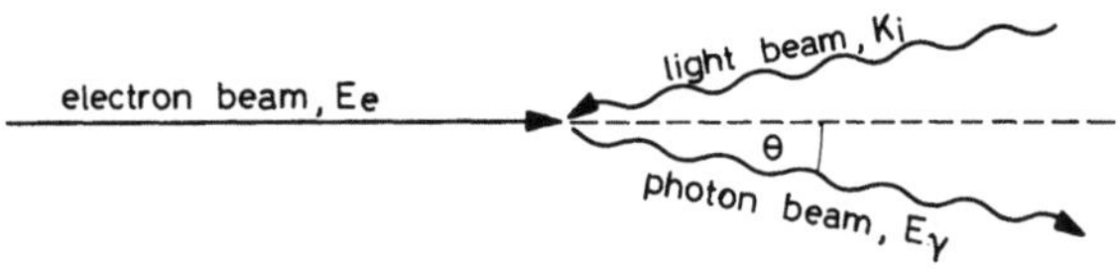

Fig. 1. Schematic diagram for the Compton backscattering of a light beam with quantum energy k_i on an electron beam of energy E_e; E_γ and θ are energy and angle of the scattered photon

Fig. 2 shows the lay out of the beam. The electron beam coming from the two-mile accelerator is brought into the interaction region by a set of bending and quadrupole magnets. The light beam is generated by a Q-switched Ruby-Laser ($\lambda = 6943$ Å) and through mirrors is directed towards the incident electrons. Backscattered photons which pass the collimator enter the 82" hydrogen bubble chamber. In order to keep the energy spread of the photon beam small the angular divergence of both the electron and the photon beam has to be less than 10^{-5} radians. This is achieved by strong collimation. The shaded distributions on bottom of Fig. 2 show the energy spectra measured. The width at half maximum is found to be $\pm 1.8\%$ at 1.46 GeV and $\pm 3.3\%$ at 4.7 GeV. The average degree of linear polarization is $\sim 94\%$.

b) Counter Experiments

There have been quite a variety of different setups in counter experiments. *Bulos et al.* [14] have used a quasi monochromatic photon beam and detected the $\pi^+\pi^-$ pairs in a set of wire chambers. The largest number of rho events was reported by the DESY-MIT group [13] which used a bremsstrahlung beam and a two-arm spectrometer to measure π^+ and π^-. The drawback of this experiment, and also of that of the Cornell group [12] which used a similar technique, lies in the fact that back-

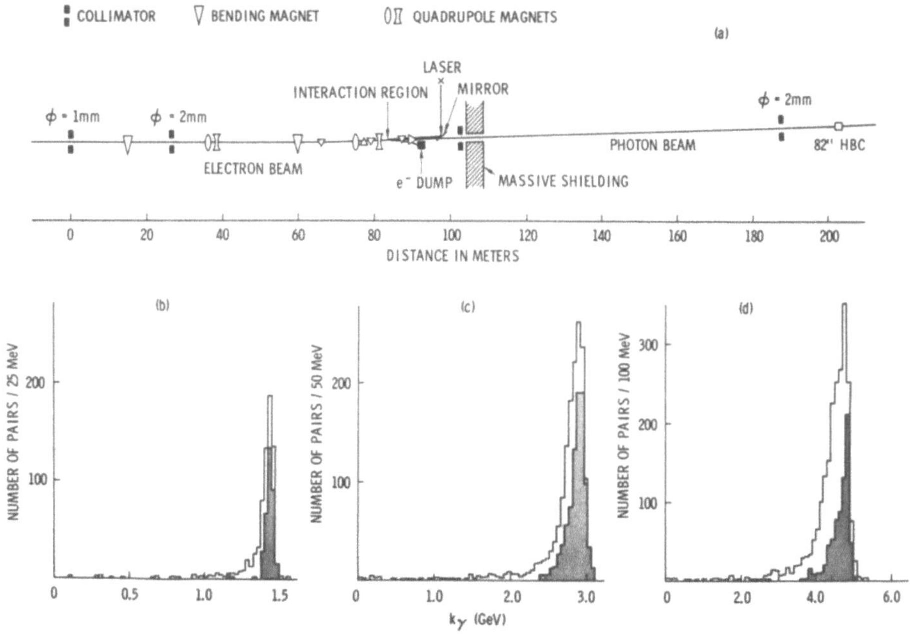

Fig. 2. a Experimental layout of the beam: b–d The pair-energy spectra at 1.4, 2.8 and 4.7 GeV. The solid histograms show the energy spectra of events fitting the reaction $\gamma p \to p\pi^+\pi^-$. (Fig. taken from Ref. [9])

ground of $\pi^+\pi^-$ pairs coming from inelastic ϱ^0 production cannot be excluded. We have used the data of the SLAC-Berkeley-Tufts (SBT) experiment in order to estimate this contamination. The events of the type

$$\gamma p \to \pi^+\pi^- + \text{anything} \tag{4}$$

were subjected to selection criteria which corresponded closely to those of the DESY-MIT experiment*. Fig. 3 shows the $\pi^+\pi^-$ mass distribution which is obtained when the events from the reaction $\gamma p \to p\pi^+\pi^-$

* In the DESY-MIT experiment the spectrometer is set to a central $\pi^+\pi^-$ momentum of $\bar{P}_{+-} = E_\gamma^{\max}/1.15$ where $E_\gamma^{\max}$ is the endpoint energy of the bremsstrahlung spectrum. The $\pi^+\pi^-$ momenta, P_{+-}, have to be within $\frac{|\bar{P}_{+-} - P_{+-}|}{\bar{P}_{+-}} \leqq 0.18$. Only a small fraction of the decay angular distribution around $\theta_H = 90°$ in the $\pi^+\pi^-$ helicity system is accepted. The total $\pi^+\pi^-$ production is obtained assuming a $\sin^2\theta_H$ distribution. It is not possible to simulate exactly these conditions with the SBT data because of the different form of the photon spectra. We have used $\bar{P}_{+-} = E_\gamma^{\max}/1.08$ in order to compensate for this difference. Also for reasons of statistics all $\pi^+\pi^-$ decay angles have been accepted in the case of the SBT data. The experimentally observed $\cos\theta_H$ distribution of the inelastic $\pi^+\pi^-$ pairs in the ϱ region is roughly constant.

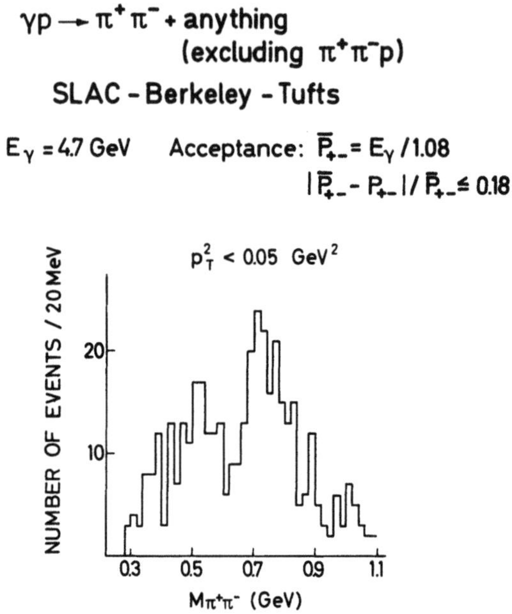

Fig. 3. Distribution of the $\pi^+\pi^-$ effective mass at $E_\gamma = 4.7$ GeV for events of the reaction $\gamma p \to \pi^+\pi^- +$ anything but excluding events of the type $\gamma p \to \pi^+\pi^- p$. Only events which satisfied the following criteria are shown: 1. the lab. momentum of the $\pi^+\pi^-$ system, P_{+-}, has to be within $|\bar{P}_{+-} - P_{+-}|/\bar{P}_{+-} \leqq 0.18$ where $\bar{P}_{+-} = E_\gamma/1.08$ and $E_\gamma = 4.7$ GeV. 2. The square of the transverse momentum, P_T^2, has to be less than 0.05 GeV^2. The data are those of the SBT-collaboration

are excluded (i.e., retaining only "background" events). Events with small transverse momenta, $P_T = P_{+-} \sin\theta_{+-}$ (θ_{+-} lab. production angle of the $\pi^+\pi^-$ system) have been selected. One concludes from Fig. 3 that there is "inelastic" $\pi^+\pi^-$ production which satisfies the acceptance criteria and, in particular, that at small P_T values there is inelastic ϱ^0 production which in the case of the Cornell and the DESY-MIT experiments will be mistaken for elastic ϱ^0 production.

In Fig. 4 the SBT-data for $\gamma p \to \pi^+\pi^- +$ anything and $\gamma p \to p\pi^+\pi^-$ are compared. The mass spectra for the total $\pi^+\pi^-$ pair production are consistently higher than those for the final state $p\pi^+\pi^-$ alone (shaded histograms). For comparison Figs. 5 and 6 show the $\pi^+\pi^-$ mass distributions measured in the Cornell and DESY-MIT experiments*. In order

* In Fig. 5 two mass spectra are shown. They were obtained by running with different endpoint energies (7.0 and 7.4 GeV) but at the same average $\pi^+\pi^-$ momentum ($\bar{P}_{+-} = 6$ GeV). The spectra are normalized to the same photon flux for elastic $\pi^+\pi^-$ production. The idea behind this exercise was that if there are inelastic contributions they will be larger in the case of the higher endpoint energy. Because the two mass spectra are nearly the same

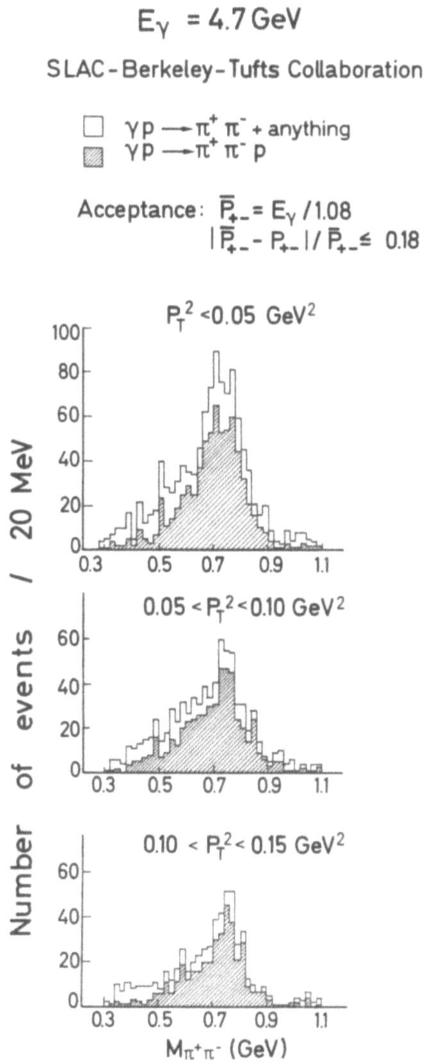

Fig. 4. Distribution of the $\pi^+\pi^-$ effective mass at $E_\gamma = 4.7$ GeV for the reaction $\gamma p \rightarrow \pi^+\pi^- +$ anything and $\gamma p \rightarrow \pi^+\pi^- p$ (shaded histograms) for different regions of the transverse momentum P_T. Only events are shown with a lab. momentum of the $\pi^+\pi^-$ system, P_{+-}, in the interval $|\bar{P}_{+-} - P_{+-}| \leqq 0.18$ where $\bar{P}_{+-} = E_\gamma/1.08$ and $E_\gamma = 4.7$ GeV

for $M_{\pi^+\pi^-} > 0.6$ GeV the authors concluded that the inelastic contributions can be neglected for $M_{\pi^+\pi^-} > 0.6$ GeV. Their test, however, is rather insensitive to inelastic contributions. The mass spectrum for the higher endpoint energy for $M_{\pi^+\pi^-} > 0.6$ GeV exceeds the other by 6%. A crude estimate shows that this could correspond to at least 20% contamination by inelastic $\pi^+\pi^-$ production.

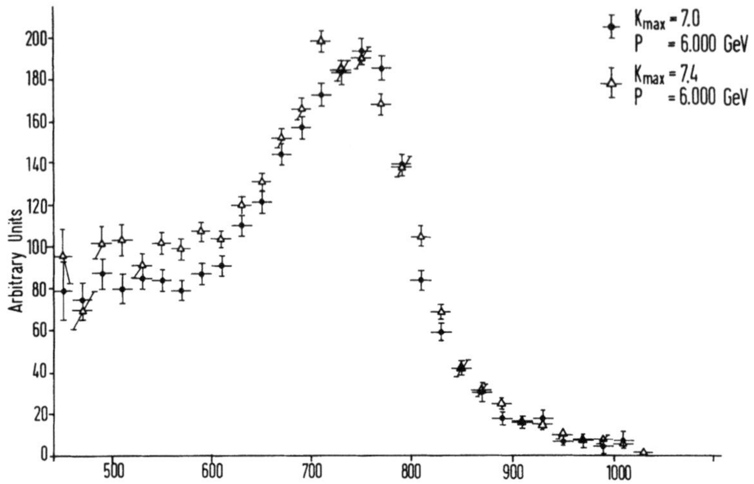

Fig. 5. Comparison of $\pi^+\pi^-$ mass spectra for photoproduction on protons for different peak bremsstrahlung energies with the same central spectrometer momentum. (Figure taken from Ref. [13])

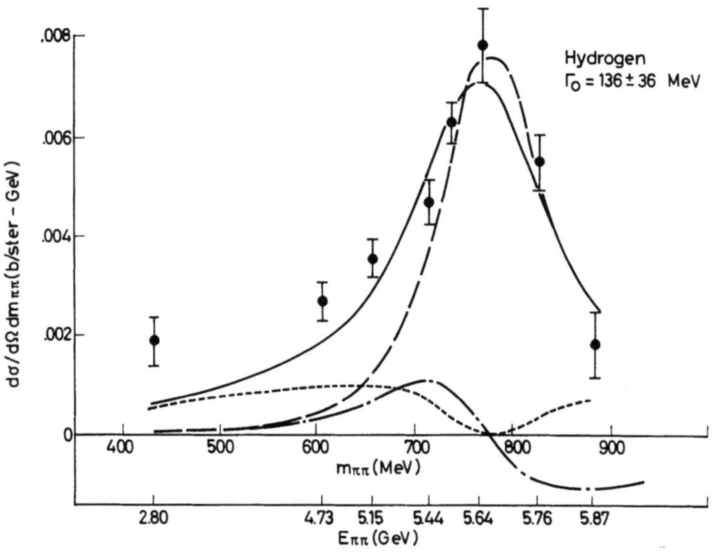

Fig. 6. The $\pi^+\pi^-$ mass spectrum for photoproduction on protons in the energy region 4–8 GeV. (Data taken from Ref. [12])

to obtain an *order of magnitude*⋆ estimate for the inelastic contributions on the DESY-MIT and Cornell experiments we define the contamination

⋆ In order to deduce from the bubble chamber data the amount of inelastic contributions seen in the two counter experiments to better than a factor of two would require a more careful analysis.

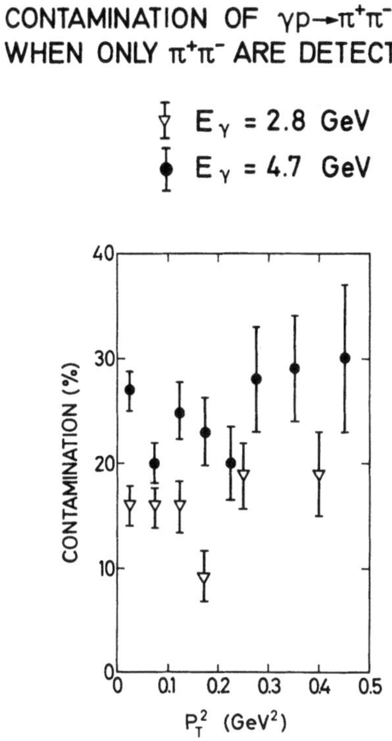

Fig. 7

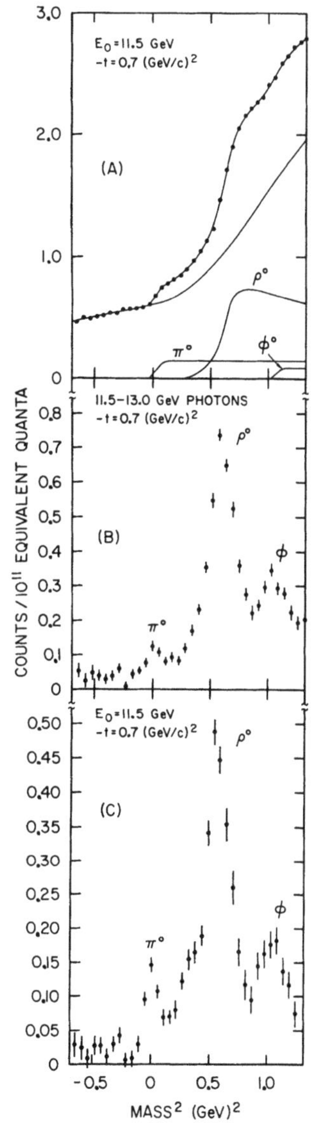

Fig. 8

Fig. 7. Contamination of $\pi^+\pi^-$ pairs from the reaction $\gamma p \to p\pi^+\pi^-$ by inelastic $\pi^+\pi^-$ production in the case that only π^+ and π^- are detected (see text). The data are those of the SBT-collaboration

Fig. 8. a The measured proton yield for a peak bremsstrahlung energy of 11.5 GeV and for $|t| = 0.7\,\mathrm{GeV}^2$. b The subtracted yield curve for photons between 11.5 and 13.0 GeV and for $|t| = 0.7\,\mathrm{GeV}^2$. c The differential yield curve derived from the integral yield curve in Fig. 8a. (Figure taken from Ref. [15])

from inelastic contributions as

$$\frac{\text{Number of inelastic } \pi^+\pi^- \text{ pairs}}{\text{Total numbers of } \pi^+\pi^- \text{ pairs}}$$

for $0.6 < M_{\pi^+\pi^-} < 0.9$ GeV and use the acceptance criteria given above. Fig. 7 shows the contamination as a function of P_T^2 for $E_\gamma = 2.8$ and 4.7 GeV. The contamination is fairly constant with P_T^2 but increases with the photon energy from $\sim 15\%$ at 2.8 GeV to $\sim 25\%$ at 4.7 GeV. At small values of P_T^2 approximately one half of the contamination is due to inelastic ϱ^0 production; i.e. $\sim 10\%$ of the forward going ϱ^0's are produced inelastically.

We conclude therefore that for a reliable measurement of $\gamma p \to p\varrho^0$ it is not sufficient to detect only π^+ and π^- when the photon energy is unknown.

Another large statistics experiment measuring rho production is that of *Anderson et al.* [15] who use a bremsstrahlung beam and detect the recoiling proton. With a missing mass technique the two-body processes $\gamma p \to p\chi^0$ ($\chi^0 = \pi^0, \varrho^0, \omega, \ldots$) are identified. The mass resolution in this experiment is not sufficient to separate ϱ^0 and ω production and to determine the rho mass shape (see the yield curve shown in Fig. 8). A 10% subtraction is applied in order to get the ϱ^0 cross section.

2. Results

Fig. 9 shows some typical $\pi^+\pi^-$ mass spectra. As is well known the photoproduced ϱ^0 has the wrong mass shape [1, 2]: contrary to a p-wave Breit-Wigner form the low mass tail of the ϱ^0 is enhanced and the high mass tail is suppressed. When the $\pi^+\pi^-$ mass distribution is fitted with a p-wave Breit-Wigner the ϱ mass appears to be shifted downwards by about 40 MeV. Whereas this problem of the ϱ mass shape causes a rather large uncertainty of the ϱ^0 production cross sections it has practically no influence on the analysis of the ϱ^0 decay angular distribution. We therefore start with the results on this subject.

a) The ϱ^0 Decay Angular Distribution

An analysis of the full ϱ^0 decay angular distribution for the case of linearly polarized photons has been reported by the SLAC-Berkeley-Tufts Collaboration for photon energies of 2.8 and 4.7 GeV [9]. For a better understanding let me give some theoretical background: *Choice of coordinate system:* In analyzing the decay distribution one has the choice between different coordinate systems, the most popular are the Gottfried-Jackson

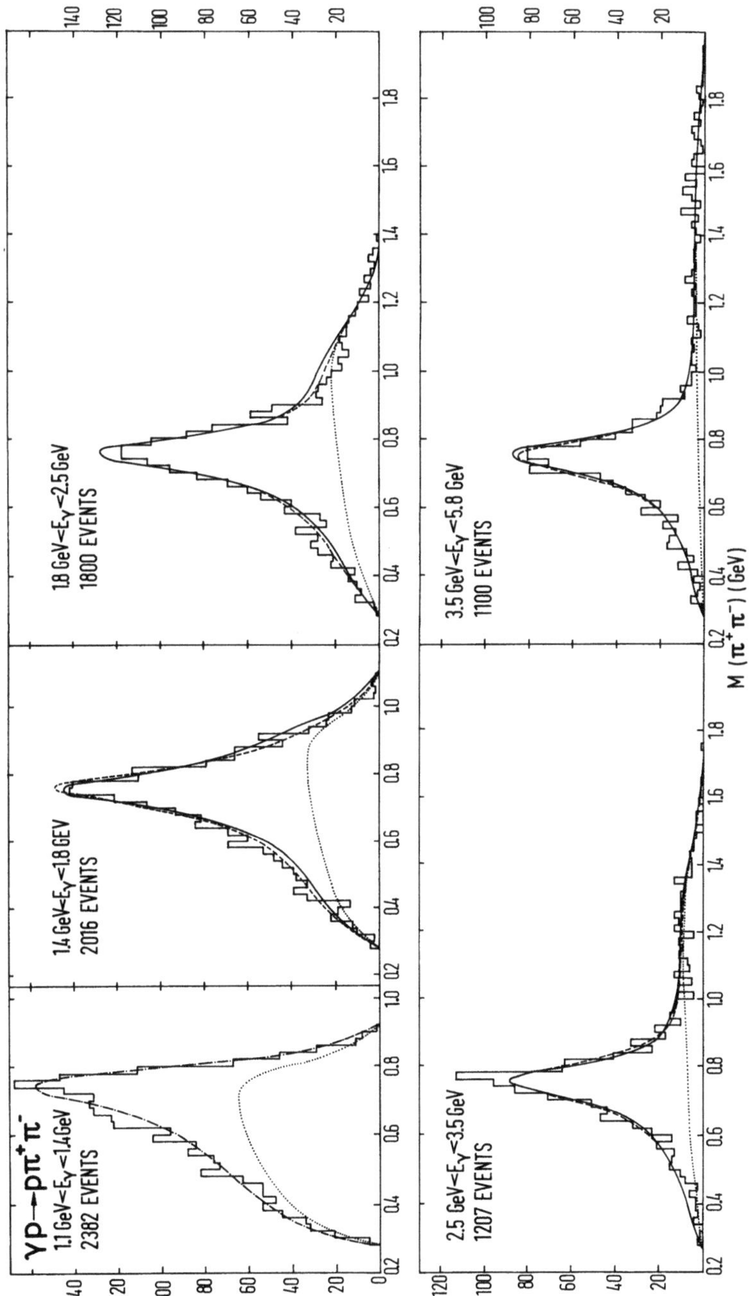

Fig. 9. Reaction $\gamma p \rightarrow p\pi^+\pi^-$. Distribution of the $\pi^+\pi^-$ mass for different intervals of the photon energy. (Figure taken from Ref. [2])

(G.-J.) system, the helicity system (H) and the Adairsystem (A). These three systems differ in the choice of the quantization axis:

G.-J.: (see Fig. 10). The quantization axis is defined as the direction of the photon in the ϱ rest system. The direction of the decay π^+ defines the polar angle θ and the azimuthal angle φ (with respect to the production plane); Φ is the angle between the photon polarization vector and the production plane. The decay distribution becomes particularily simple if we introduce the polarization angle $\Psi = \varphi - \Phi$.

H.: The ϱ direction in the total c.m.s. defines the quantization axis.

A.: The photon direction in the total c.m.s. defines the quantization axis.

Which of the three coordinate systems is to be preferred depends on the production mechanism. This is illustrated in Fig. 11 where a circularly polarized photon is considered and where σ_z^γ, σ_z^ϱ indicate the directions along which the spins of γ and ϱ have components equal to one. The G.-J. system is preferred when the t-channel helicity is conserved (e.g. spin 0^+ exchange). The H.-system is preferred when the s-channel helicity is conserved in the c.m.s. whereas A. is the appropriate system when the spin is conserved in the s-channel c.m.s. (spin independence). For forward production all three systems coincide.

For all three systems the ϱ^0 decay angular distribution expressed in terms of the rho density matrix, ϱ_{ik}, is of the same form [16]:

$$\begin{aligned} W(\cos\theta, \varphi, \Phi) = {} & W^0(\cos\theta, \varphi, \varrho_{ik}^0) - P_\gamma \cos 2\Phi\, W^1(\cos\theta, \varphi, \varrho_{ik}^1) \\ & - P_\gamma \sin 2\Phi\, W^2(\cos\theta, \varphi, \varrho_{ik}^2) \\ = {} & \frac{3}{4\pi}\left[\frac{1}{2}(1-\varrho_{00}^0) + \frac{1}{2}(3\varrho_{00}^0 - 1)\cos^2\theta - \sqrt{2}\,\mathrm{Re}\,\varrho_{10}^0 \sin 2\theta \cos\varphi \right. \\ & - \varrho_{1-1}^0 \sin^2\theta \cos 2\varphi - P_\gamma \cos 2\Phi \left\{\varrho_{11}^1 \sin^2\theta + \varrho_{00}^1 \cos^2\theta \right. \\ & \left. - \sqrt{2}\,\mathrm{Re}\,\varrho_{10}^1 \sin 2\theta \cos\varphi - \varrho_{1-1}^1 \sin^2\theta \cos 2\varphi \right\} \\ & \left. - P_\gamma \sin 2\Phi \left\{\sqrt{2}\,\mathrm{Im}\,\varrho_{10}^2 \sin 2\theta \sin\varphi + \mathrm{Im}\,\varrho_{1-1}^2 \sin^2\theta \sin 2\varphi \right\}\right] \end{aligned} \tag{5}$$

where P_γ is the degree of linear polarization and the density matrix of the rho has been split into three parts:

$$\varrho_{ik} = \varrho_{ik}^0 - P_\gamma \cos 2\Phi\, \varrho_{ik}^1 - P_\gamma \sin 2\Phi\, \varrho_{ik}^2 . \tag{6}$$

The first term, namely $W^0(\cos\theta, \varphi, \varrho_{ik}^0)$ describes the unpolarized case. With linearly polarized photons one can measure nine independent density matrix elements in contrast to the three elements obtainable with an unpolarized beam.

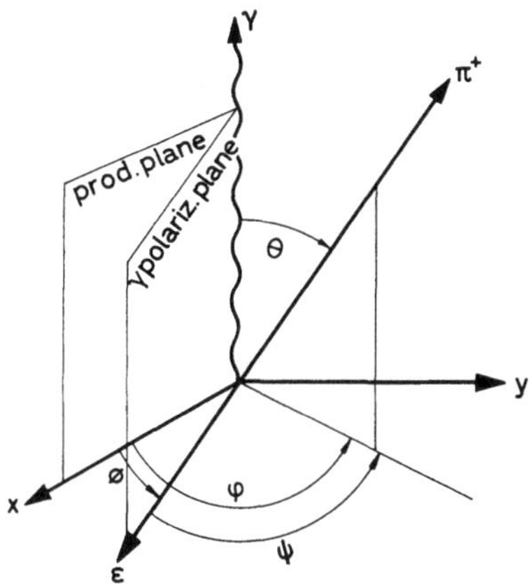

Fig. 10. Analysis of the ϱ^0 decay angular distribution: the Gottfried-Jackson system in the ϱ^0 rest frame

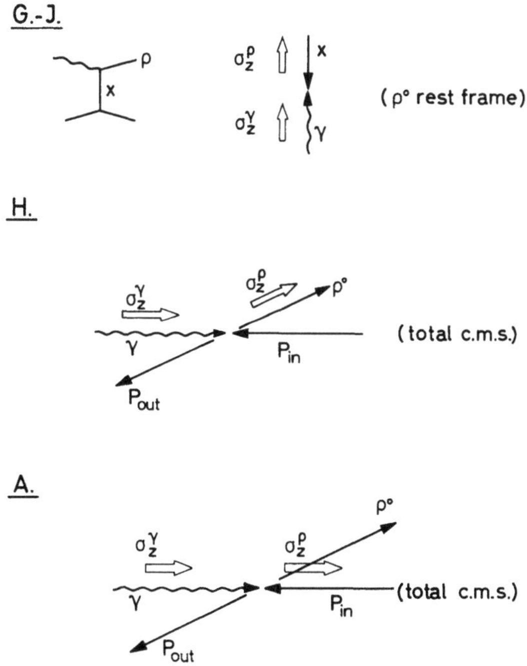

Fig. 11. Illustration of the spin dependence of some production mechanisms for $\gamma p \to p \varrho^0$ which lead to a simple behaviour of the rho density matrix in either the Gottfried-Jackson (G.-J.), the helicity (H) or the Adair (A) frame; the photon is assumed to be circularly polarized; σ_z^γ, σ_z^ϱ indicate the directions along which the spins of γ and ϱ have components equal to one

As a special case we give the density matrix of a linearly polarized photon beam:

$$\varrho^1_{1-1} = -\operatorname{Im}\varrho^2_{1-1} = 0.5,$$

all other ϱ^α_{ik} which appear in Eq. (5) equal to zero.

For the following discussion it is useful to remember that the density matrix elements can be written as a sum of bilinear combinations of helicity amplitudes:

$$\varrho_{\lambda_\varrho \lambda'_\varrho} = \sum_{\lambda_\gamma, \lambda''_\gamma} T_{\lambda_\varrho \lambda_\gamma} T^*_{\lambda'_\varrho \lambda''_\gamma}, \qquad \lambda_\gamma, \lambda''_\gamma = 1, -1 \tag{7}$$

(the sum over the nucleon helicities has already been carried out). As an example

$$\varrho^0_{00} \sim |T_{01}|^2 + |T_{0-1}|^2$$

measures only helicity flip amplitudes, i.e. if $\varrho^0_{00} \neq 0$ we know that there are helicity flip contributions.

One of the advantages of studying vector meson production with linearly polarized photons is that it allows one to measure separately the cross section contributions σ^{N} and σ^{U} from the exchange of natural $(P=(-1)^J)$ and unnatural parity $(P=-(-1)^J)$ in the t-channel. In particular one can measure the parity asymmetry, P_σ,

$$P_\sigma = \frac{\sigma^{\mathrm{N}} - \sigma^{\mathrm{U}}}{\sigma^{\mathrm{N}} + \sigma^{\mathrm{U}}}$$

which at high energies in terms of the vector meson density matrix is given by [16]

$$P_\sigma = 2\varrho^1_{1-1} - \varrho^1_{00}. \tag{9}$$

Roughly speaking, natural and unnatural-party exchange contributions manifest themselves in the decay angular distribution as follows: For natural-parity exchange the decay pions emerge preferrentially in the plane of photon polarization ($\Psi \cong 0$), and for unnatural-parity exchange perpendicular to it ($\Psi \simeq 90°$).

In some experiments the probability for emission of pions from ϱ-decay is only measured at $\theta = 90°$, $\varphi = 90°$ at the two points: $\Psi = 0°$ $(\sigma_\parallel)$ and $\Psi = 90° (\sigma_\perp)$, leading to the asymmetry [16]

$$\Sigma = \frac{\sigma_\parallel - \sigma_\perp}{\sigma_\parallel + \sigma_\perp} = \frac{\varrho^1_\parallel + \varrho^1_{1-1}}{\varrho^0_\parallel + \varrho^0_{1-1}}. \tag{10}$$

The quantity Σ is closely related to P_σ; it measures the contributions from natural and unnatural-parity exchange in the t-channel to the transverse vector meson states.

Now we have the tools to look at the ϱ decay angular distributions. Fig. 12 shows in the helicity frame the distributions of $\cos\theta$ and Ψ for events in the ϱ region at 2.8 and 4.7 GeV. The polar angular distribution is proportional to $\sin^2\theta$ which implies that the ϱ mesons are produced with helicity ± 1 in the total c.m.s. The Ψ distributions show the effect of the photon and the rho polarization: the events cluster around $\Psi = 0°$ and $180°$; only a few events are observed near $\Psi = 90°$ and $270°$. The foregoing discussion then tells us that natural-parity exchange dominates.

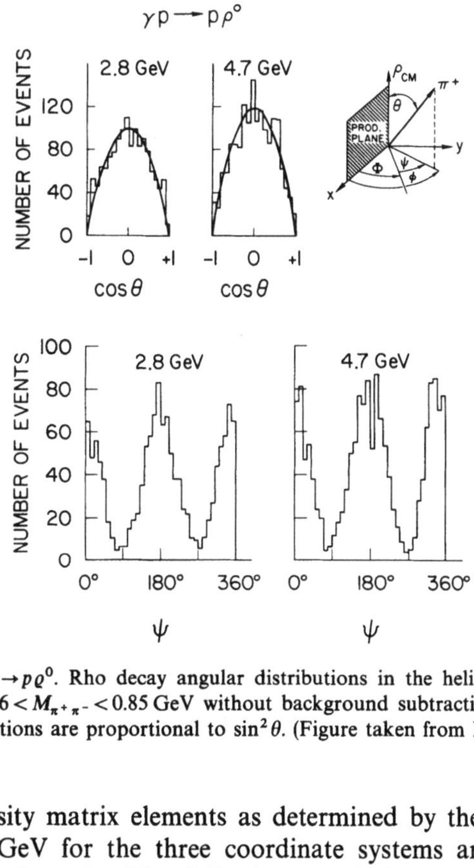

Fig. 12. Reaction $\gamma p \to p\varrho^0$. Rho decay angular distributions in the helicity system for $|t| < 0.4\ \mathrm{GeV}^2$ and $0.6 < M_{\pi^+\pi^-} < 0.85$ GeV without background subtraction. The curves for the $\cos\theta$ distributions are proportional to $\sin^2\theta$. (Figure taken from Ref. [9])

The rho density matrix elements as determined by the SBT-collaboration at 4.7 GeV for the three coordinate systems are shown in Fig. 13 as a function of t. The following conclusions can be drawn:

1. The ϱ^{α}_{ik} vary rapidly in the G.-J. frame, e.g., ϱ^0_{00} which receives contributions from t-channel helicity flip amplitudes only, increases rapidly with increasing momentum transfer; therefore *t-channel helicity is not conserved.* This result was already known from the CEA bubble chamber experiment [1].

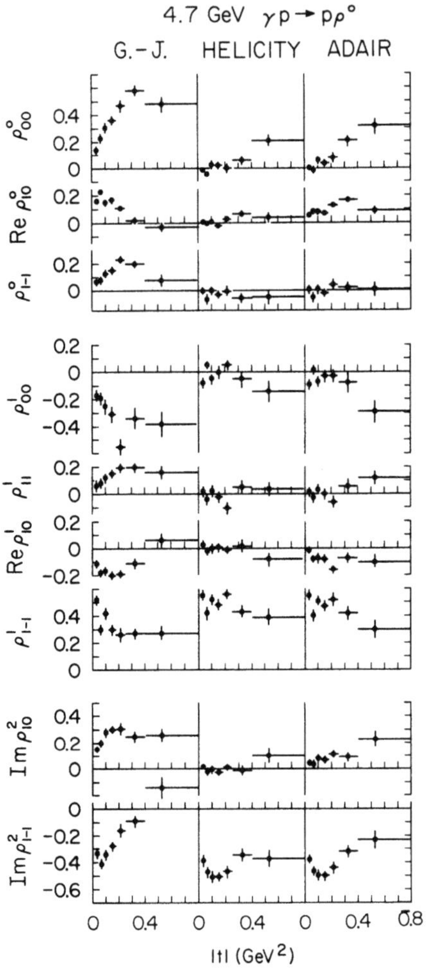

Fig. 13. Reaction $\gamma p \rightarrow p \varrho^0$. The spin density matrix parameters as a function of t in the Gottfried-Jackson, helicity and Adair systems. (Figure taken from Ref. [9])

2. The ϱ^{α}_{ik} vary also in the Adair system (see also Fig. 14a), although the variation there is not as dramatic. This variation *excludes* the hypothesis of spin independence in the s-channel c.m. system [17].

3. In the helicity system the ϱ^{α}_{ik} show practically no t-dependence up to $|t| \lesssim 0.4\,\mathrm{GeV}^2$. The helicity flip terms are zero within errors.

The fact that the flip terms are minimum in the helicity system is further demonstrated in Figs. 14b and c. Here the rho density matrix as measured in the helicity frame was rotated around the normal to the

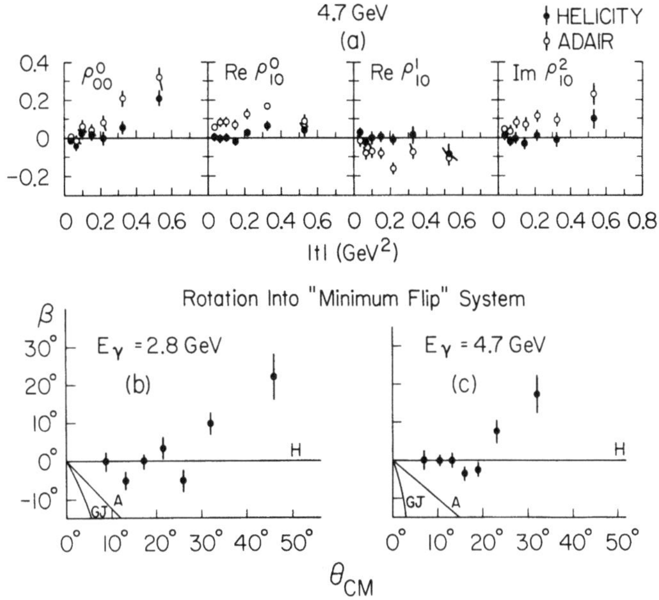

Fig. 14a–c. Reaction $\gamma p \to p \varrho^0$. a Comparison of the spin density matrix parameters ϱ^0_{00}, $\mathrm{Re}\,\varrho^0_{10}$, $\mathrm{Im}\,\varrho^2_{10}$ in the helicity (♦) and Adair systems (◊) as a function of t (data of Fig. 13). b, c The angle β for rotation into the "minimum flip" system as a function of the c.m.s. rho production angle θ_{cm}. For the curves labeled G.J., A, and H, see text. (Figure taken from Ref. [9])

production plane by an angle β and a least squares fit was made to find that value of β for which the flip terms become minimal. Figs. 14b and c show β as a function of the c.m.s. production angle, θ^*, along with curves which indicate the expected behaviour of β if the flip terms are minimal in the G.-J., A or H systems. For $\theta^* \lesssim 25°$ (corresponding to $|t| \lesssim 0.4\,\mathrm{GeV}^2$) the flip terms are minimal in the helicity system.

We conclude that in the c.m.s. the rho in reaction (1) behaves like a proton: the spin is aligned along the ϱ direction of flight. To state the result differently: Rho photoproduction *conserves helicity in the s-channel total c.m. system* up to $|t| \leqq 0.4\,\mathrm{GeV}^2$. Combining this result with the evidence found by *Höhler* and *Strauss* [5] for πN scattering *Gilman et al.* [18] have suggested that diffraction processes *in general* conserve the helicity in the s-channel c.m. system.

Fig. 15 shows the parity asymmetry, P_σ, as determined at 2.8 and 4.7 GeV from the measured values of the ϱ^α_{ik}. At both energies P_σ is close to unity, i.e., as anticipated already from the Ψ distributions (Fig. 12) rho production is dominated by natural-party exchange. From the value of P_σ averaged over $|t| \leqq 1\,\mathrm{GeV}^2$ one finds for the unnatural-

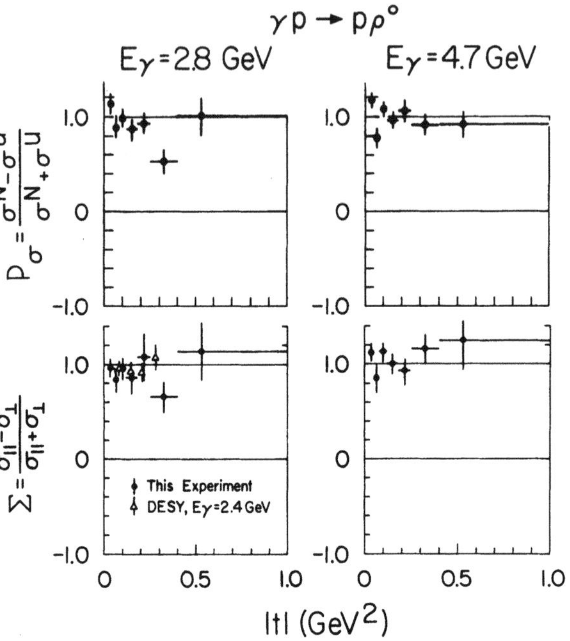

Fig. 15. Reaction $\gamma p \to p \varrho^0$. The parity asymmetry, P_σ, and the asymmetry, Σ, as a function of t in the Gottfried-Jackson, helicity and Adair systems. (Figure taken from Ref. [9])

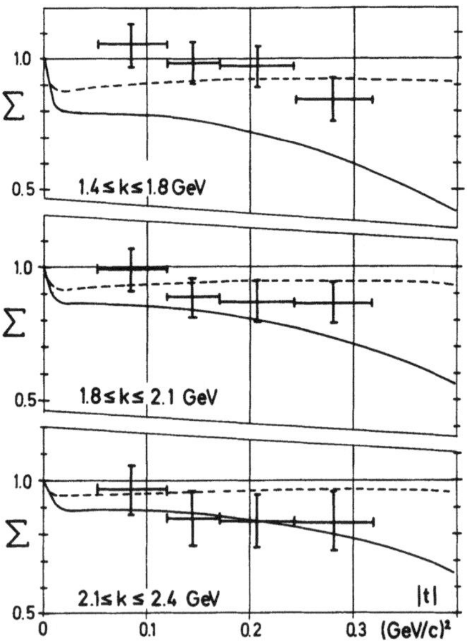

Fig. 16. Asymmetry Σ for different regions of the photon energy. (Figure taken from Ref. [19])

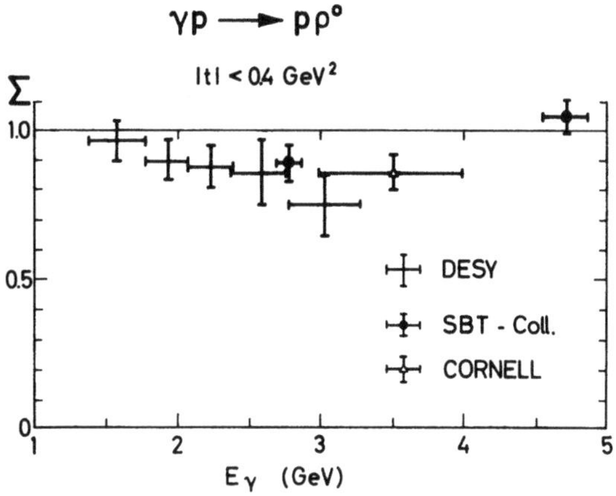

Fig. 17. Energy dependence of the asymmetry Σ for $|t| < 0.4\,\mathrm{GeV}^2$. (Data taken from Refs. [9, 19 and 20])

parity contribution

$$\sigma^{\mathrm{U}}/\sigma^{\mathrm{tot}} \equiv \tfrac{1}{2}(1-P_\sigma) = 3.1 \pm 2.2\% \quad \text{at } 2.8\,\mathrm{GeV} \text{ and } (-1.1 \pm 1.9)\% \text{ at } 4.7\,\mathrm{GeV}.$$

From one-pion exchange alone we expect $\sigma^{\mathrm{U}}/\sigma^{\mathrm{tot}}$ to be $\sim 2\%$ at 2.8 GeV and 0.8% at 4.7 GeV.

The dominance of natural-parity exchange is also evident from the behaviour of the asymmetry Σ. *Criegee et al.* [19] have measured the t dependence of Σ for photon energies between 1.4 and 2.4 GeV (see Fig. 16). The energy dependence of Σ for $|t| < 0.4\,\mathrm{GeV}^2$ is given in Fig. 17. In all cases and even near threshold Σ is found to be close to one*.

b) *Cross Section for ϱ^0 Production*

As mentioned before the determination of cross sections for ϱ^0 production is complicated by the fact that the ϱ^0 mass shape is not fully understood yet. As a consequence the ϱ^0 cross sections deduced from the data depend on the specific model used. The uncertainty in the ϱ^0 cross sections caused by the theoretical difficulties are of the order of 20–30%. There exists, however, a quantity related to the ϱ^0 cross section which can be measured in a model independent way: the intensity Π for the production of $\pi^+\pi^-$ pairs in the s-channel helicity conserving p-wave state [9]. This then constitutes an upper limit for the cross section of ϱ^0 production through s-channel helicity conservation. The decay angular distribution in this case (with $P_\gamma = 1$) is given by (see

* In the experiment of Ref. [19], however, Σ was determined using *all* pairs in the ϱ^0 region.

above):

$$W(\cos\theta, \Psi) = \frac{3}{4\pi}\sin^2\theta\cos^2\Psi$$
$$= \frac{3}{8\pi}\{\sin^2\theta + \sin^2\theta\cos 2\Psi\}. \tag{11}$$

The second term is proportional to the spherical harmonic Re $Y_2^2(\theta, \Psi)$ and can be measured through moment analysis leading directly to the

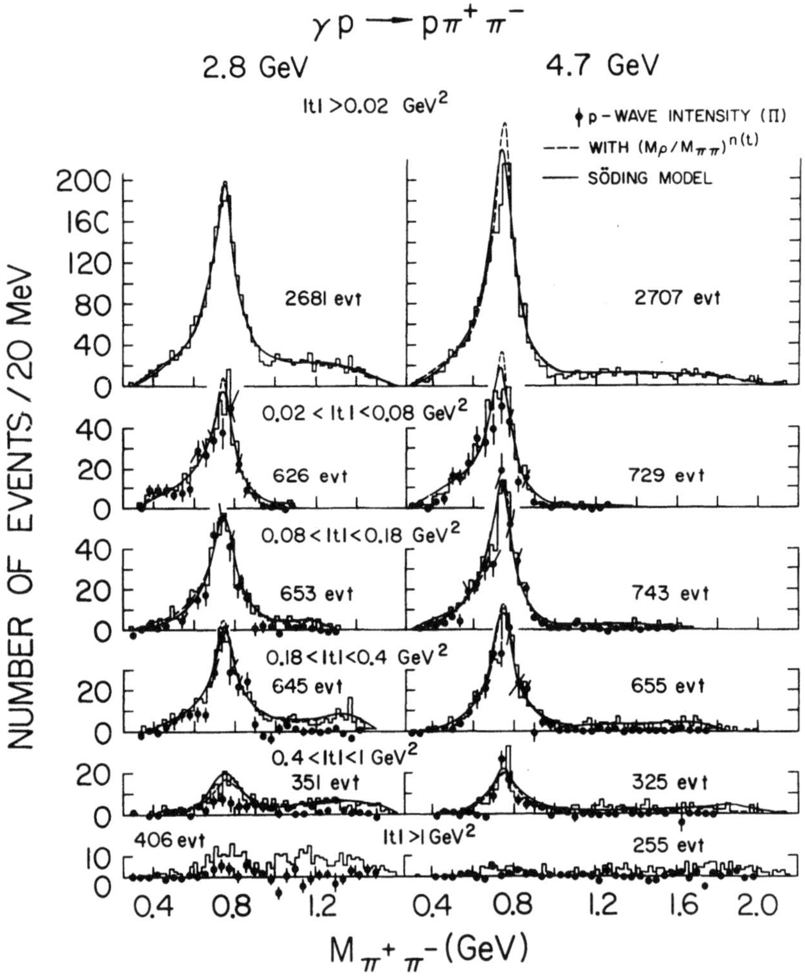

Fig. 18. $\pi^+\pi^-$ mass distributions for events of the reaction $\gamma p \to p\pi^+\pi^-$. The helicity-conserving p-wave intensity Π (see text) is shown by the points ⧫. The curves give the results of a maximum likelihood fit using for the ϱ^0 the parametrization $(M_\varrho/M_{\pi\pi})^{n(t)}$ (- - -) and the Söding model (——). (Figure taken from Ref. [9])

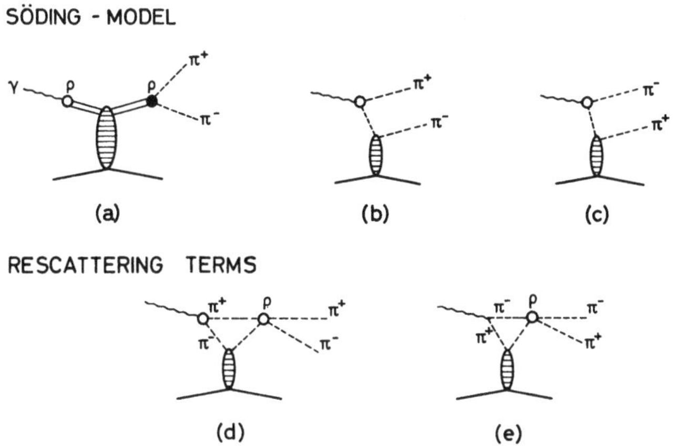

Fig. 19a–e. Diagrams for the production of $\pi^+\pi^-$ pairs in the reaction $\gamma p \to p\pi^+\pi^-$. a ϱ^0 production; b, c Drell processes; d, e rescattering terms

intensity Π:

$$\Pi = \sqrt{40\,\pi/3}\sum_i \mathrm{Re}\, Y_2^2(\theta_i, \Psi_i) \tag{12}$$

where the summation has to be done over all events. Fig. 18 shows the $\pi^+\pi^-$ mass spectra as measured at 2.8 and 4.7 GeV by the SBT-collaboration for different regions of t. The dots mark the values of Π. Except for $|t| > 1\ \mathrm{GeV}^2$ or $M_{\pi^+\pi^-} > 1$ GeV Π is close to the total $\pi^+\pi^-$ mass distribution which implies that nearly all $\pi^+\pi^-$ pairs are produced in a helicity conserving p-wave state. One also notices that Π shows the same mass skewing as do the distributions for all events.

We now turn to the ϱ^0 production cross section and discuss some of the models [21–27] which try to explain the ϱ mass shift:

1. According to *Ross* and *Stodolsky* [21] near $t = 0$ the rho Breit-Wigner has to be multiplied by $(M_\varrho/M_{\pi^+\pi^-})^4$, a factor which stems from the photon propagator. This factor causes a skewing of the ϱ mass shape.

2. In the Söding model [22] the ϱ mass shift is due to an interference of the ϱ^0 production amplitude with a Drell-type background (see diagrams a–c of Fig. 19). In order to avoid double counting corrections *Yennie* [23] and *Pumplin* [24] have suggested to add the rescattering terms illustrated by the diagrams d, e in Fig. 19. As a result the p-wave contributions from the Drell process (diagrams b, c) are multiplied by the factor

$$\frac{M_{\pi^+\pi^-}^2 - M_\varrho^2}{M_{\pi^+\pi^-}^2 - M_\varrho^2 + i M_\varrho \Gamma_\varrho}$$

i.e. these contributions vanish for $M_{\pi^+\pi^-} = M_\varrho$.

3. *Kramer* and coauthors [25] have taken a point of view just opposite to that of *Söding*. They consider the rescattering terms (diagrams d, e of Fig. 19), which in the Söding model represent a small correction, as the only contributions to ϱ^0 production. In this model all p-wave $\pi\pi$ production is called ϱ^0 production and therefore Π represents the intensity for s-channel helicity conserving ϱ^0 production.

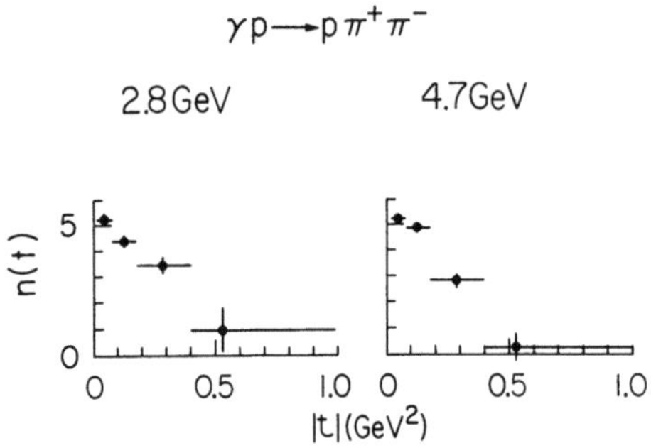

Fig. 20. Reaction $\gamma p \to p\pi^+\pi^-$: Fitted values for $n(t)$ using the parametrization $(M_\varrho/M_{\pi^+\pi^-})^{n(t)}$. (Figure taken from Ref. [9])

The SBT-collaboration has tested the Ross-Stodolsky factor by multiplying the Breit-Wigner for the rho with the factor $(M_\varrho/M_{\pi^+\pi^-})^{n(t)}$ and fitting for the exponent n in different t intervals. This parametrization describes the $\pi^+\pi^-$ mass spectra well (see Fig. 18). The resulting values of $n(t)$, as shown in Fig. 20, depend strongly on t: n is close to 5 near $t=0$ and zero above $|t| \simeq 0.4$ GeV^2, i.e. the ϱ mass shift disappears with increasing $|t|$. The conclusion then is that the Ross-Stodolsky factor should not be used at larger $|t|$ values.

The SBT-collaboration has also analyzed the reaction $\gamma p \to p\pi^+\pi^-$ in terms of the Söding model. In the fits the amount of Drell background was treated as a free parameter. The model was found to reproduce the data well with the fitted amount of Drell background being close to that calculated directly. The solid curves for the $\pi^+\pi^-$ mass spectra in Fig. 18 were obtained with the Söding model.

The differential cross sections determined with the Söding model for $|t| \lesssim 0.4$ GeV^2 are considerably *smaller* than the values obtained for $\Pi(t)$ or those resulting from the parametrization $(M_\varrho/M_{\pi\pi})^{n(t)}$ (see Fig. 21). It also leads to a *smaller* exponential slope for the t distribution of ϱ^0 production.

Differential cross sections for ϱ^0 production obtained at higher photon energies by *Anderson et al.* [15] with the missing mass technique and

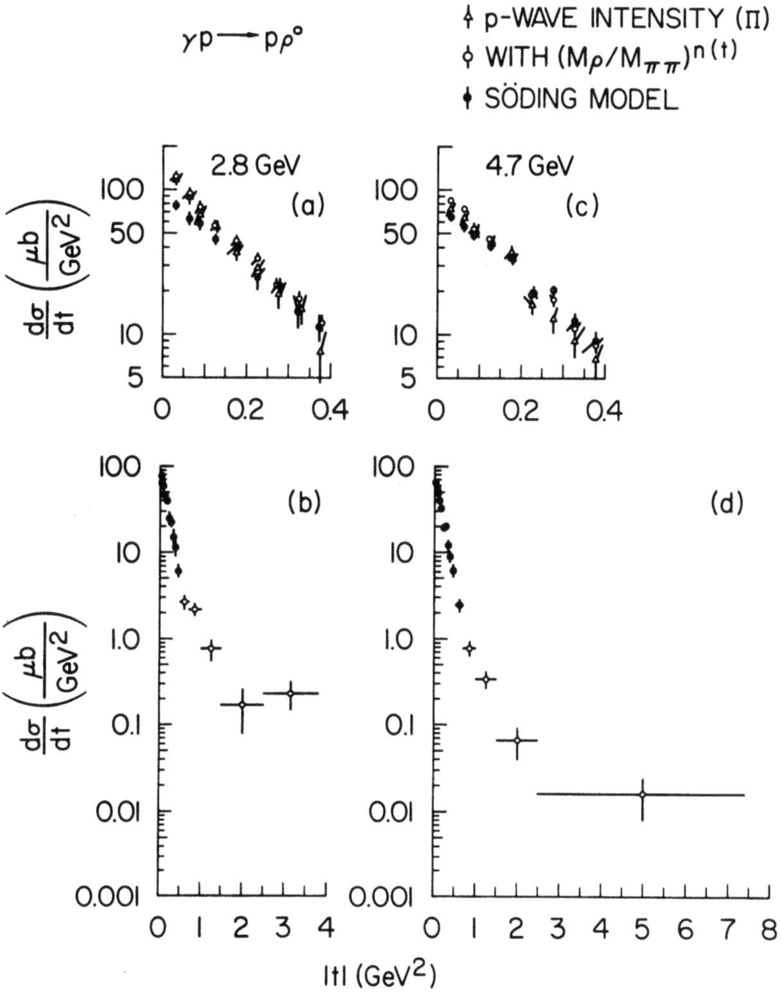

Fig. 21. Reaction $\gamma p \to p\pi^+\pi^-$: Differential cross sections as a function of t for the helicity-conserving p-wave contribution Π (♁), for ϱ^0 production as obtained from fits with the Söding model (♦) and using the parametrization $(M_\varrho/M_{\pi\pi})^{n(t)}$ (♁). (Figure taken from Ref. [9])

using the Ross-Stodolsky factor are shown in Fig. 22. Since these measurements were done above $|t| = 0.1$ GeV^2 the values of the forward differential cross section deduced from these data depend critically on how the extrapolation is made (for more details see Ref. 15). The curves shown are proportional to the expression

$$\left(\sqrt{d\sigma/dt(\pi^- p \to \pi^- p)} + \sqrt{d\sigma/dt(\pi^+ p \to \pi^+ p)}\right)^2$$

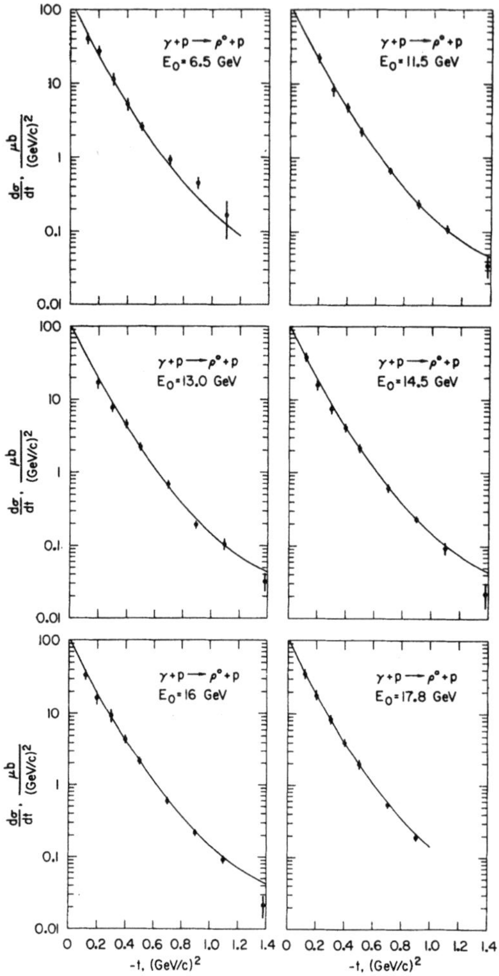

Fig. 22. Differential cross section for $\gamma p \to p \varrho^0$ for photon energies between 6.5 and 17.8 GeV. (Figure taken from Ref. [15])

suggested by the quark model. These curves follow the data points rather well. It is not clear, however, what would happen if the ϱ^0 cross sections were determined with the Söding model. In any event, these data show that the ϱ^0 differential cross section does not depend strongly on the photon energy up to 18 GeV.

In Table 2 the ϱ^0 cross sections measured in various experiments and with different fitting methods are compared. Those cross section values determined in the two bubble chamber experiments with the same fitting

Table 2. *Results on ϱ^0 production. The forward differential cross sections $\mathrm{d}\sigma^0/\mathrm{d}t$ and the slope A were obtained from a fit of $\mathrm{d}\sigma/\mathrm{d}t$ to the form $\mathrm{d}\sigma^0/\mathrm{d}t\,\exp(At)$*

Experiment	E_γ (GeV)	Fit method	M_ϱ (MeV)	Γ_ϱ (MeV)	σ (μb)	$\mathrm{d}\sigma^0/\mathrm{d}t$ (μb/GeV²)	A (GeV⁻²)
SBT	2.8	$(M_\varrho/M_{\pi\pi})^4$	765 ± 3	143 ± 7	21.4 ± 0.6	153 ± 10	6.9 ± 0.3
Ref. [9]		$(M_\varrho/M_{\pi\pi})^{n(t)}$	765 ± 3	140 ± 6	20.9 ± 1.0	143 ± 10	6.7 ± 0.3
		Söding	765 ± 3	132 ± 6	16.4 ± 1.0	93 ± 7	5.4 ± 0.4
		Söding + rescatt.	767 ± 3	131 ± 7	17.0 ± 1.0	101 ± 7	5.6 ± 0.4
		$\Pi(t)$			18.9 ± 1.1	144 ± 13	7.5 ± 0.6
DESY-HBC	2.7				21.3 ± 2.1		
	2.9				19.7 ± 2.0		
	3.5	$(M_\varrho/M_{\pi\pi})^4$			17.6 ± 0.8		
Ref. [2]	4.5				17.1 ± 0.9		
	2.5–3.5					147 ± 13	6.9 ± 0.4
	3.5–4.5					149 ± 19	8.1 ± 0.7
	2.7				17.6 ± 1.7		
	2.9	Söding			16.8 ± 1.8		
	3.5				15.2 ± 0.8		
	4.5				15.6 ± 0.9		
SBT	4.7	$(M_\varrho/M_{\pi\pi})^4$	760 ± 3	148 ± 5	15.5 ± 0.4	110 ± 8*	7.1 ± 0.3
Ref. [9]		$(M_\varrho/M_{\pi\pi})^{n(t)}$	759 ± 2	136 ± 5	15.8 ± 0.8	108 ± 6*	7.0 ± 0.3
		Söding	760 ± 3	141 ± 5	14.4 ± 0.7	80 ± 5*	5.5 ± 0.3
		Söding + rescatt.	765 ± 3	131 ± 5	15.1 ± 0.7	90 ± 6*	5.8 ± 0.3
		$\Pi(t)$			13.5 ± 1.0	102 ± 8*	7.2 ± 0.5
DESY-HBC	5.4	$(M_\varrho/M_{\pi\pi})^4$			15.6 ± 2.2		
Ref. [2]	4.5–5.8					130 ± 16	7.9 ± 0.7
	5.4	Söding			14.3 ± 2.0		
SLAC	9	$(M_\varrho/M_{\pi\pi})^4$	760 ± 10	135 ± 10		122 ± 12	
(*Bulos et al.*)		Söding				104 ± 12	
Refs. [14, 28]							

* Note added in proof: Corrections for scanning losses have increased these values to 116 ± 8, 114 ± 6, 84 ± 5, 94 ± 6 and 109 ± 8 respectively.

method agree with each other to within 1.2 s.d. The most striking differences due to the theoretical uncertainties of how to describe ϱ^0 production occur at low energies: at 2.8 GeV, the forward differential cross section obtained for $\Pi(t)$ is 144 ± 13 μb/GeV² whereas the Söding model plus rescattering corrections yields 101 ± 7 μb/GeV². One notices also that the forward differential cross sections determined with the Söding model are less energy dependent.

The forward differential cross sections deduced from the bubble chamber data depend *critically* on the assumption that $\mathrm{d}\sigma/\mathrm{d}t$ in the t range $|t|\leqq 0.4\,\mathrm{GeV}^2$ behaves like $\exp(At)$ with A being a constant. The data in $0.02 \leqq |t| \leqq 0.4\,\mathrm{GeV}^2$ are well described by this form but an increase of A for, say, $|t| \leqq 0.05\,\mathrm{GeV}^2$ and hence an increase of the forward differential cross section (by $\approx 10\%$) cannot be excluded.

Summary of rho photoproduction.

1. For a reliable measurement of $\gamma p \to p \varrho^0$ it is not sufficient to detect only π^+ and π^- when the photon energy is unknown because of the contamination by inelastic ϱ^0 production.

2. The rho mass shape is skewed leading to an apparent shift of the rho mass to lower values. This mass shift disappears with increasing momentum.

3. The problem of the rho mass shift is not fully understood yet and causes an uncertainty in the forward differential cross section for rho production, $d\sigma^0/dt$, of the order of 20–30%. With our present knowledge one expects the actual value of $d\sigma^0/dt$ to lie between that determined with the Söding model (including rescattering corrections) and the intensity Π for the production of $\pi^+\pi^-$ pairs in the s-channel helicity conserving p-wave state. Despite this problem one can conclude that the differential cross section for ϱ^0 production is approximately constant between 5 and 18 GeV.

4. Rho production proceeds almost completely through natural-parity exchange. The contributions from unnatural-parity exchange are of the order of a few percent.

5. The rho production mechanism conserves the helicity in the s-channel total c.m. system for $|t| < 0.4\,\mathrm{GeV}^2$.

III. ω Photoproduction via $\gamma p \to p \omega$

The presently available data on this reaction come all from track-chamber experiments (see Table 3). The statistical accuracy is rather limited: a total of ~ 1300 events have been observed between 2 and 16 GeV.

Table 3. *Experiments measuring* $\gamma p \to p \omega$

	technique	beam	photon energy (GeV)	number of ω events above 2 GeV
CEA-HBC [2]	HBC	bremsstrahlung	<6	100
DESY-HBC [2]	HBC	bremsstrahlung	<6	400
SLAC [30]	HBC	quasi monochromatic	5.3	30
Weizmann [8]	HBC	quasi monochromatic	4.3	30
SLAC-Berkeley-Tufts [31]	HBC	Laser beam, approx. monochrom. lin. pol.	2.8, 4.7	700
SLAC [10]	Streamer-chamber	bremsstrahlung	2–16	30

The reaction $\gamma p \to p\omega$ is detected in the final state $p\pi^+\pi^-\pi^0$. Fig. 23 shows two $\pi^+\pi^-\pi^0$ mass spectra. The ω signal stands out clearly. Because of the narrow width of the ω there is no problem in separating off the background. The energy dependence of the total cross section for ω production, σ_ω, is given in Fig. 24: σ_ω rises steeply above threshold

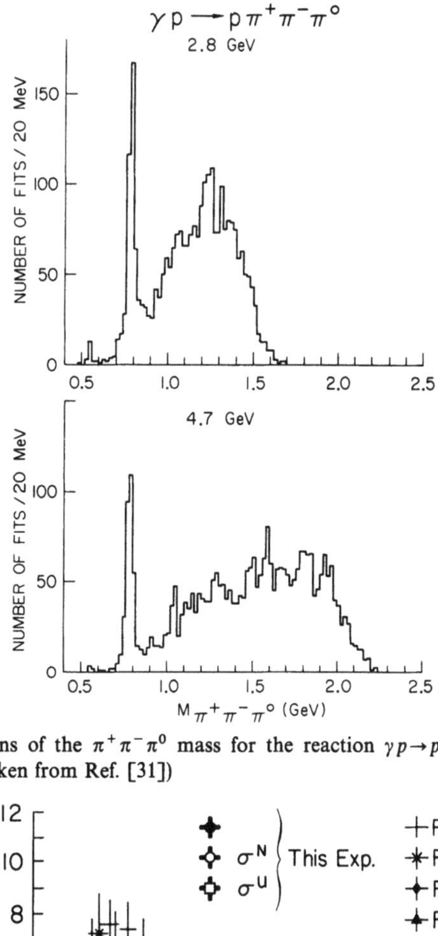

Fig. 23. Distributions of the $\pi^+\pi^-\pi^0$ mass for the reaction $\gamma p \to p\pi^+\pi^-\pi^0$ at 2.8 and 4.7 GeV. (Figure taken from Ref. [31])

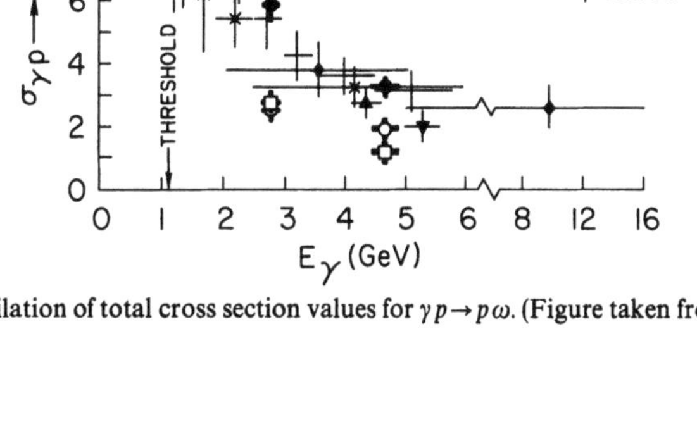

Fig. 24. Compilation of total cross section values for $\gamma p \to p\omega$. (Figure taken from Ref. [31])

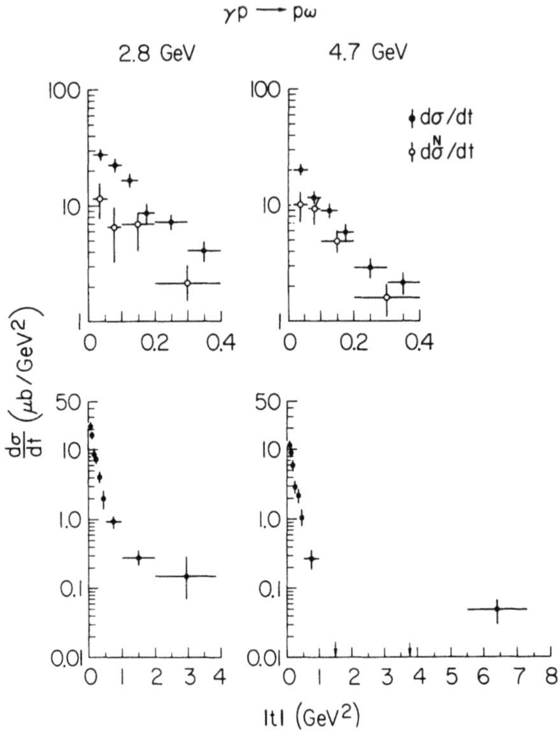

Fig. 25. Reaction $\gamma p \to p\omega$. Total differential cross sections and differential cross sections for contributions from natural-parity exchange at 2.8 and 4.7 GeV. (Figure taken from Ref. [31])

to a maximum of $\sim 8\,\mu b$ around 1.8 GeV and then, in contrast to the ϱ cross section, drops off. The fall off of the differential cross section for ω production (see Fig. 25) on the other hand, is close to that found for ϱ^0 production.

The most likely explanation of the energy dependence of σ_ω is that two production mechanism contribute, namely one-pion exchange (OPE) and diffration scattering. The first dominates the low energy region and the second takes over at higher energies. This conjecture can be tested using linearly polarized photons. As in the case of ϱ^0 production, by analyzing the ω decay angular distribution, one can separate the contributions σ^N, σ^U from natural-parity exchange (e.g. diffraction scattering) and unnatural-parity exchange (e.g. OPE). In Fig. 26 the distribution of the polarization angle Ψ is shown as measured by the SBT-collaboration at 2.8 GeV [31]. (The definition of Ψ is the same as in the case of rho mesons if the π^+ from the decay of the rho is replaced by the normal

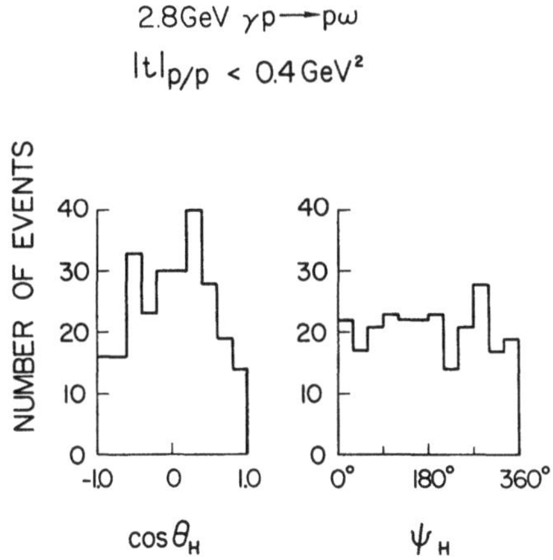

Fig. 26. Reaction $\gamma p \to p\omega$. Omega decay angular distributions in the helicity system for $|t| < 0.4\ \mathrm{GeV}^2$. (Data taken from Ref. [31])

Table 4. *Parameter of ω production. Cross sections and momentum transfer dependence for* $0.02 < |t| < 0.4\ \mathrm{GeV}^2$ *assuming* $d\sigma/dt = (d\sigma^0/dt)\exp(At)$ *for all events, and for the contributions from natural-parity exchange in the t-channel. (Data from Ref.* [31]*)*

E_γ (GeV)	σ_{total} (μb)	$d\sigma^0/dt$ (μb/GeV²)	A (GeV⁻²)	$\sigma^N(\lvert t\rvert<1\ \mathrm{GeV}^2)$ (μb)	$\sigma^U(\lvert t\rvert<1\ \mathrm{GeV}^2)$ (μb)	$d\sigma^{0N}/dt$ (μb/GeV²)	A^N (GeV⁻²)
2.8	5.8±0.5	34±4	6.2±0.7	2.50±0.37	2.70±0.39	13.1±4.1	5.5±1.6
4.7	3.2±0.3	25±3	8.0±0.8	1.84±0.28	1.25±0.27	15.2±3.8	7.5±1.5

to the ω decay plane.) In striking contrast to the corresponding distribution for rho production the Ψ distribution for ω events is flat which implies that at this energy both, natural and unnatural-parity exchange contribute with approximately equal strength. The values deduced for σ^N, σ^U at 2.8 and 4.7 GeV are given in Fig. 24 and in Table 4. The natural parity cross section does not show a significant energy dependence whereas σ^U decreases from 2.8 to 4.7 GeV. The energy dependence and the magnitude of σ^U agree with the predictions for OPE. The t dependence of σ^N is shown in Fig. 25; the values obtained from these data for the slope and the forward cross section are listed in Table 4.

In summary, the available data on $\gamma p \to p\omega$ are consistent with the picture that this reaction proceeds via one-pion exchange and via a diffraction mechanism. There may be, however, a large positive contribution (20–40%) to σ^N from A_2 exchange and from the interference of A_2 with P and P' exchange [32].

IV. φ Production via $\gamma p \to \varphi p$

The data on this reaction are still rather scanty due to the small production cross section; e.g. the total number of φ events from published track chamber experiments is less than 50. The largest sample of φ events has been observed by *Anderson et al.* [15] using the missing mass technique mentioned before. However this experiment covers the region $|t| \gtrsim 0.3\,\text{GeV}^2$ which contains only $\sim 20\%$ of the total φ production, and therefore prevents an accurate determination of the forward differential cross section. At DESY [33] and at *Cornell* [34] this reaction has also been studied; in these experiments a contamination from inelastic φ production cannot be excluded. Results from the bubble chamber experiment at DESY [35] indicate that the inelastic contributions could be of comparable magnitude: For energies between 4.5 and 5.8 GeV the cross section for $\gamma p \to p \pi^+ \pi^- \varphi$ was found to be $0.6 \pm 0.2\,\mu\text{b}$.

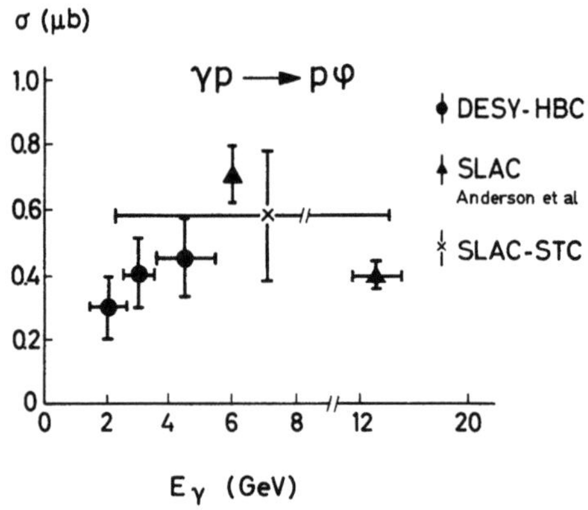

Fig. 27. Compilation of total cross section values for $\gamma p \to p\varphi$. (Data taken from Refs. [2, 10 and 15])

A compilation of the total cross section for φ production is shown in Fig 27; σ_φ is probably energy independent and of the order of $0.5\,\mu\text{b}$. The measurements by *Anderson et al.* [15] of the differential cross section are given in Fig. 28.

To conclude the section on φ production let me remind you that – as *Freund* [36] has pointed out several years ago – the reaction $\gamma p \to p\varphi$ is best suited for a study of the pomeron trajectory. The measurements to be done are obvious: accurate determination of the energy and t dependence of the cross section and of the φ decay angular distribution, preferrably with linearly polarized photons.

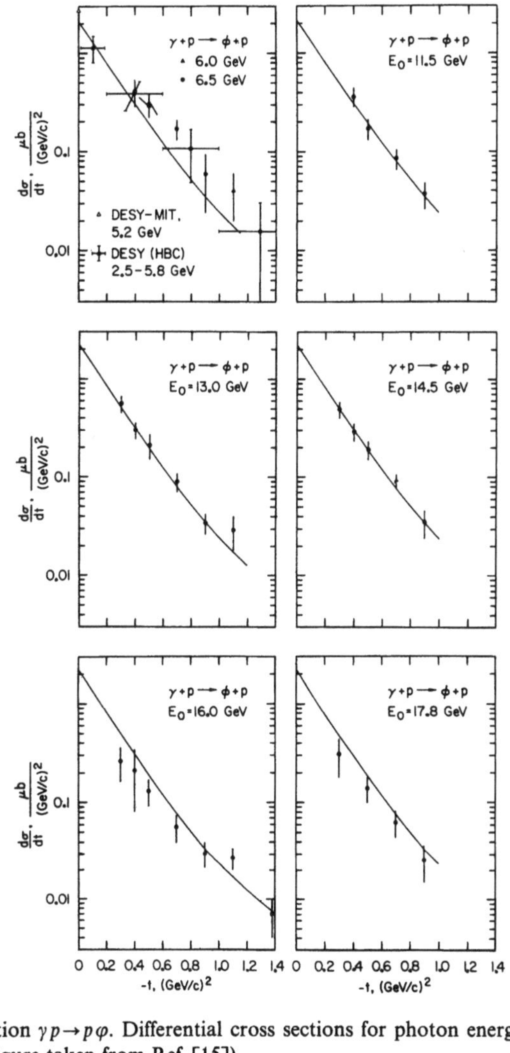

Fig. 28. Reaction $\gamma p \to p\varphi$. Differential cross sections for photon energies between 6 and 17.8 GeV. (Figure taken from Ref. [15])

V. Comparison with the Vector Dominance Model

The vector dominance model (VDM) relates the photoproduction of vector mesons, $\gamma p \to pV$, ($V = \varrho^0, \omega, \varphi$) to the total hadronic photon proton cross section, $\sigma_T(\gamma p)$, and to vector meson production through electron-positron annihilation, $e^+e^- \to V$. The starting point for all VDM relations is the conjecture that the electromagnetic current, j_μ, can be written as a sum of the vector meson fields [6]:

$$j_\mu(x) = -\left\{\frac{M_\varrho^2}{2\gamma_\varrho}\varrho_\mu(x) + \frac{M_\omega^2}{2\gamma_\omega}\omega_\mu(k) + \frac{M_\varphi^2}{2\gamma_\varphi}\varphi_\mu(x)\right\} \tag{13}$$

where M_V is the mass of the vector meson and γ_V measures the strength of the photon-vector meson coupling. There exists an extensive literature on the application of Eq. (13); see e.g., the review papers quoted under Ref. [37].

We shall consider the following three VDM relations:

1.
$$\begin{aligned}\left.\frac{d\sigma}{dt}\right|_{t=0}(\gamma p\to Vp) &= \frac{\alpha/4}{\gamma_V^2/4\pi}\left.\frac{d\sigma}{dt}\right|_{t=0}(Vp\to Vp)\\ &= \frac{\alpha/64\pi}{\gamma_V^2/4\pi}\,\sigma_T^2(Vp).\end{aligned} \tag{14}$$

In deriving Eq. (14) the following assumptions were made:

a) the scattering is purely diffractive;

b) the scattering amplitude is the same for different helicity states;

c) contributions of the type $V'p\to Vp(V'\neq V)$ can be neglected.

With the predictions of SU_6 [38],

$$1/\gamma_\varrho^2 : 1/\gamma_\omega^2 : 1/\gamma_\varphi^2 = 9:1:2^\star \tag{15}$$

and of the quark model [37a, 41]

$$\begin{aligned}&\sigma_T(\varrho p) = \sigma_T(\omega p) = \tfrac{1}{2}(\sigma_T(\pi^+ p) + \sigma_T(\pi^- p)) \simeq 28\text{ mb at } 5\text{ GeV}\\ &\sigma_T(\varphi p) = \sigma_T(k^+ p) + \sigma_T(k^- p) - \sigma_T(\pi^+ p) \simeq 13\text{ mb at } 5\text{ GeV}\end{aligned} \tag{16}$$

one obtains for the *diffractive* part of vector meson production

$$\frac{d\sigma}{dt}(\gamma p\to\varrho^0 p) : \frac{d\sigma}{dt}(\gamma p\to\omega p) : \frac{d\sigma}{dt}(\gamma p\to\varphi p) = 9:1:0.4 \tag{17}$$

Experimentally one finds for this ratio**

at 2.8 GeV: $9:(1.2\pm0.4):(0.40\pm0.08)$

at 4.7 GeV: $9:(1.5\pm0.4):(0.45\pm0.09)$.

The experimental data are compatible with the prediction (17).

2. We can use the relation (14) together with Eq. (16) and determine $\gamma_\varrho^2/4\pi$ from the measured values of $d\sigma/dt(\gamma p\to\varrho^0 p)$ given in Table 2.

*Modifications due to symmetry breaking change this ratio either to 9:0.65:1.33 [39] or to 9:1.21:1.34 [40].

** For ϱ^0 and ω production the values of Refs. [9, 31] were taken, using for ϱ^0 the cross sections obtained with the Söding model including rescattering corrections and for ω only the natural-parity exchange contributions. The φ data were taken from Ref. [2] and are an average over $2.5 < E_\gamma < 5.8$ GeV.

This leads to the values for $\gamma_\varrho^2/4\pi$ listed in Table 5. The SBT-data yield values for $\gamma_\varrho^2/4\pi$ which are larger than the value obtained in the Orsay storage ring experiment, which is $\gamma_\varrho^2/4\pi = 0.50 \pm 0.05$ or 0.54 ± 0.05 (depending on the assumptions made for the behaviour of the ϱ width [42]).

3. The most direct test of VDM is provided by the following relation, which involves no other assumption:

$$\sigma_T(\gamma p) = \sum_{V=\varrho,\omega,\varphi} \sqrt{\frac{4\pi\alpha}{\gamma_V^2/4\pi} \frac{d\sigma}{dt} \mathrm{diff}(\gamma p \to Vp)}. \tag{18}$$

Table 5. *Determination of* $\gamma_\varrho^2/4\pi$ *using* VDM *(Eq.* (14)*) and the quark model (Eq.* (16)*)*

E_γ	Ref.	Fit method for $d\sigma^0/dt(\gamma p \to \varrho^0 p)$	$\gamma_\varrho^2/4\pi$
2.8	[9]	Söding + rescattering	0.89 ± 0.06
		$\Pi(t)$	0.63 ± 0.06
4.7	[9]	Söding + rescattering	0.82 ± 0.06
		$\Pi(t)$	0.71 ± 0.06
9	[14, 28]	Söding	0.60 ± 0.07

Table 6. *Determination of* $\gamma_\varrho^2/4\pi$ *from Eq.* (18)

Input $\frac{d\sigma^{0\,(\mathrm{diff})}}{dt}(\gamma p \to \omega p) = 14 \pm 5\ \mu b/\mathrm{GeV}^2$ *(see Table 4)*

$\frac{d\sigma^{0\,(\mathrm{diff})}}{dt}(\gamma p \to p) = 3.6 \pm 1\ \mu b/\mathrm{GeV}^2$ *(Ref.* [2]*)*

$\frac{\gamma_\omega^2}{4\pi} = 3.5 \pm 0.7 \quad \frac{\gamma_\varphi^2}{4\pi} = 2.5 \pm 0.23$

(from the Orsay storage ring experiment, Ref. [42]*) Eq.* (18) *then leads to*

$$\gamma_\varrho^2/4\pi = \frac{36 \frac{d\sigma^{0\,(\mathrm{diff})}}{dt}(\gamma p \to \varrho^0 p)}{(\sigma_T(\gamma p) - 18.6 \pm 2.7)^2}$$

with $d\sigma/dt$ *and* σ_T *in units of* $\mu b/\mathrm{GeV}^2$ *and* μb respectively

E_γ (GeV)	Fit method	$\frac{d\sigma^0}{dt}(\gamma p \to \varrho^0 p)$ ($\mu b/\mathrm{GeV}^2$)	$\sigma_T(\gamma p)$ (μb)	$\gamma_\varrho^2/4\pi$
2.8	Söding + rescattering	101 ± 7 [9]	131 ± 4 [43, 44]	0.29 ± 0.04
	$\Pi(t)$	144 ± 13 [9]		0.41 ± 0.06
4.7	Söding + rescattering	90 ± 6 [9]	124 ± 4 [43, 44]	0.29 ± 0.04
	$\Pi(t)$	102 ± 8 [9]		0.33 ± 0.04
9	Söding	104 ± 12 [14, 28]	120 ± 4 [45]	0.36 ± 0.06

Here the diffractive part of the forward cross sections for $\gamma p \to Vp$ are to be inserted. Since on the r.h.s. of Eq. (18) the by far largest contribution comes from the ϱ we will use Eq. (18) to determine $\gamma_\varrho^2/4\pi$. Table 6 lists the input values used and gives the result. The large uncertainties of the ω and φ quantities have only a minor effect on the determination of $\gamma_\varrho^2/4\pi$. Also, in calculating $\gamma_\varrho^2/4\pi$, the assumption was made that ϱ^0 production is purely diffractive. Dispersion calculations on the other hand suggest that the ratio of real to imaginary part is of the order of 20% [46]. Taking this into account will reduce the values of $\gamma_\varrho^2/4\pi$ shown in Table 6 by ~4%. Table 6 shows that the value of $\gamma_\varrho^2/4\pi$ obtained from the VDM relation (19) and the measured data on $d\sigma/dt(\gamma p \to Vp)$ and $\sigma_T(\gamma p)$ is *smaller* than the storage ring result*.

A similar conclusion has recently been reached from a comparison of Compton scattering with the photoproduction of vector mesons using Eq. (18) [47]. The result of *Buschhorn et al.* is shown in Fig. 29.

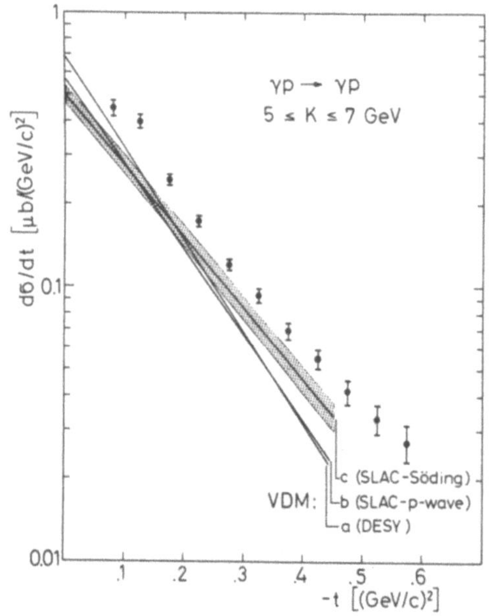

Fig. 29a–c. Measured Compton cross sections and VDM predictions. The systematic error of 7% is not included in the experimental points. a ϱ^0 from DESY-HBC; b ϱ^0 from SLAC-HBC: *s*-channel helicity conserving *p*-wave; c ϱ^0 from SLAC-HBC: Söding model. Error region for a and b similar as for c. (Figure taken from *Buschhorn et al.*, Ref. [47])

* One should keep in mind, however, that the values of the forward differential ϱ^0 cross section at 2.8 and 4.7 GeV were obtained assuming the form $\exp(At)$ for all $|t| \leqq 0.4\,\mathrm{GeV}^2$ (see section II).

We conclude that for photoproduction of vector mesons VDM in its present form works only to within a factor of ~ 1.5.

Finally, we briefly discuss some of the corrections proposed to bring the vector-dominance model into agreement with experiment:

1. The strength of the photon vector meson coupling, i.e. γ_V, depends on the mass of the photon, M_γ. In the reaction $e^+e^- \to V$ one has $M_\gamma = M_V$ whereas in the photoproduction case, $M_\gamma = 0$.

2. The scattering amplitude for $VN \to VN$ depends on M_V [48]. This is an assumption similar to that made by *Zichichi* [49] in a model of the nucleon form factor where the VNN vertex depends on M_V.

3. There exist additional vector mesons which couple to the photon and which have to be included in the r.h.s. of Eq. (13). Although no such state has been observed yet [50] the present experiments cannot exclude the existence of a $T=1$ vector meson, ϱ', with the following characteristics:

$M_{\varrho'} = 1\text{–}2\ \text{GeV}$

$\Gamma_{\varrho'} \cong 200\ \text{MeV}$

decay predominantly *not* into $\pi^+\pi^-$

$\sigma(\gamma p \to \varrho' p) \lesssim 0.5\ \mu\text{b}$.

The existence of such a ϱ' could explain why the relation (18) leads to a smaller value for $\gamma_\varrho^2/4\pi$. Furthermore, if the $\gamma\varrho$ and $\gamma\varrho'$ coupling constants have opposite sign (e.g. as in the Veneziano ansatz of *Di Vecchia* and *Drago* [51] for the nucleon form factor) we can understand why the VDM relation (14) yields larger values for $\gamma_\varrho^2/4\pi$ when applied

1. to ϱ^0 photoproduction on protons together with the quark model prediction Eq. (16) (see Table 5);

2. to ϱ^0 photoproduction on nuclei. The effect of a ϱ' contribution in connection with the latter reaction has been discussed e.g. by *Cordes* and *O'Donnell* [52] and by *Leith* [14].

Acknowledgements. I am very grateful to Profs. *Busza* and *Lohrmann* for comments and criticism.

References

1. *CEA-Bubble-Chamber-Group.* Phys. Rev. **146**, 994 (1966); – *Lanzerotti, L. J., et al.*: Phys. Rev. Letters **15**, 210 (1965).
2. *Aachen-Berlin-Bonn-Hamburg-Heidelberg-München* Collaboration. Phys. Rev. **175**, 1669 (1968).
3. *Blechschmidt, H., et al.*: Proceedings of the International Symposium on Electron and Photon Interactions at High Energies (1965), Deutsche Physikalische Gesellschaft, Hanau, Germany, (1966), Vol. II, p. 173.
4. *Berman, S. M., Drell, S. D.*: Phys. Rev. **133**, B 791 (1964).
5. *Höhler, G., Strauss, R.*: Z. Physik **232**, 205 (1969).

6. *Sakurai, J. J.:* Ann. Phys. **11**, 1 (1960); – *Gell-Mann, M., Zachariasen, F.:* Phys. Rev. **124**, 953 (1961); – *Nambu, Y., Sakurai, J. J.:* Phys. Rev. Letters **8**, 79 (1962); – *Gell-Mann, M., Sharp, D., Wagner, W. G.:* Phys. Rev. Letters **8**, 261 (1962).
7. *Ballam, J., et al.:* Phys. Rev. Letters **21**, 1541 (1968).
8. *Eisenberg, Y., et al.:* Phys. Rev. Letters **22**, 669 (1969).
9. *SLAC-Berkeley-Tufts Collaboration.* Phys. Rev. Letters **23**, 498 (1969); **24**, 955 (1970); **24**, 960 (1970).
10. *Davier, M., et al.:* Phys. Rev. D **1**, 790 (1970).
11. *Blechschmidt, H., et al.:* Nuovo Cimento **52** A, 1348 (1967).
12. *McClellan, G., et al.:* Phys. Rev. Letters **22**, 374 (1969) and preprint (1970).
13. *Alvensleben, H., et al.:* Phys. Rev. Letters **23**, 1058 (1969).
14. *Bulos, F., et al.:* see data quoted by *Leith, D. W. G. S.:* SLAC-Report No. SLAC-PUB-679 (1969).
15. *Anderson, R., et al.:* Phys. Rev. D **1**, 27 (1970).
16. See e.g.: *Schilling, K., Seyboth, P., Wolf, G.:* Nucl. Phys. B **15**, 397 (1970) – and References given therin.
17. This assumption was made in the "strong absorption model", *Eisenberg, Y., et al.:* Phys. Letters **22**, 217 (1966); **22**, 223 (1966); see also *Kramer, G.:* DESY Report No. 67/32, (1967) (unpublished).
18. *Gilman, F. J., Pumplin, J., Schwimmer, A., Stodolsky, L.:* Phys. Letters **31** B, 387 (1970).
19. *Criegee, L., et al.:* Phys. Rev. Letters **25**, 1306 (1970).
20. *Diambrini-Palazzi, G., et al.:* Phys. Rev. Letters **25**, 478 (1970).
21. *Ross, M., Stodolsky, L.:* Phys. Rev. **149**, 1172 (1966).
22. *Söding, P.:* Phys. Letters **19**, 702 (1966); see also *Krass, A.:* Phys. Rev. **159**, 1496 (1967).
23. *Yennie, D.:* Private communication and *Bauer, T.:* Phys. Rev. Letters **25**, 485 (1970).
24. *Pumplin, J.:* SLAC-Report No. SLAC-PUB-730 (1970).
25. *Kramer, G., Uretsky, J. L.:* Phys. Rev. **181**, 1918 (1969); – *Kramer, G., Quinn, H. R.:* DESY Report No. DESY 70/23 (1970); see also: *Kramer, G.:* Internal Report No. DESY T-70/4 (1970).
26. *Satz, H., Schilling, K.:* CERN Report No. TH-1127 (Jan. 1970).
27. *Mannheim, P. D., Maor, U.:* Preprint (1970), to be published.
28. *Leith, D. W. G. S.:* Private communication.
29. *CEA-Bubble Chamber Group.* Phys. Rev. **155**, 1468 (1967).
30. *Ballam, J., et al.:* Phys. Letters **30** B, 421 (1969).
31. *SLAC-Berkeley-Tufts Collaboration.* Phys. Rev. Letters **24**, 1364 (1970).
32. *Harari, H.:* Proceedings of the 4th International Symposium on Electron and Photon Interactions at High Energies, Liverpool, England (1969), (ed. by *D. W. Braben*), p. 107.
33. *DESY-MIT group.* Work quoted in Ref. [37 b].
34. *Mistry, N. B., et al.:* Work quoted by A. Silverman, Proceedings of the 4th International Symposium on Electron and Photon Interactions at High Energies, Liverpool, England (1969), (ed. by *D. W. Braben*), p. 71.
35. *Aachen-Berlin-Bonn-Hamburg-Heidelberg-München Collaboration.* Phys. Letters **27** B, 54 (1968) and Phys. Rev. **188**, 2060 (1969).
36. *Freund, P. G. O.:* Nuovo Cimento **48**, 541 (1967); see also: *Barger, V., Cline, D.:* Phys. Rev. Letters **24**, 1313 (1970).
37a. *Joos, H.:* Proceedings of the International Conference on Elementary Particles, Heidelberg, Germany (1967), (ed. by *H. Filthuth*), p. 349.
37b. *Ting, S. C. C.:* Proceedings of the 14th Conference on High Energy Physics, Vienna, Austria (1968), (ed. by *J. Prentki* and *J. Steinberger*), p. 43;
37c. *Lohrmann, E.:* Proceedings of the Lund International Conference on Elementary Particles held at Lund, Sweden, 25 June–1 July 1969 (Berlingska Bortryskeriet, Lund, Sweden, 1970), p. 13;

37d *Sakurai, J. J.*: Proceedings of the International Symposium on Electron and Photon Interactions at High Energies, Liverpool, England (1969), (ed. by *D. W. Braben*), p. 91;
37e. *Stodolsky, L.*: Orsay-Report No. 69/51 (1969);
37f. *Diebold, R.*: Talk given at the High Energy Physics Conference, Boulder, Colorado, USA (1969), SLAC-Report No. SLAC-PUB-673 (1969);
37g. *Guiragossian, Z. G. T.*: Talk given at the International Seminar on Vector Mesons and Electromagnetic Interactions, Dubna, U.S.S.R., Dubna (1969), SLAC-Report No. SLAC-PUB-694 (1969);
37h. *Schildknecht, D.*: Talk given at the V-th "Rencontre de Moriond", Méribel-les-Allues, France (1970), DESY Report No. DESY 70/31 (1970).
38. *Sakita, B.*, *Wali, K. C.*: Phys. Rev. **139**, B 1355 (1965).
39. *Oakes, R. J.*, *Sakurai, J. J.*: Phys. Rev. Letters **19**, 1266 (1967).
40. *Das, T.*, *Mathur, V. S.*, *Okubo, S.*: Phys. Rev. Letters **19**, 470 (1967).
41. *Krammer, M.*: DESY Report No. DESY 70/53 (1970). The total $\pi^{\pm} p$ cross sections were taken from the compilation of: *Giacomelli, G.*: CERN Report No. CERN-HERA 69-3 (1969). The KN cross sections were taken from: *Galbraith, W., et al.*: Phys. Rev. B **138**, 913 (1965).
42. See: *Perez-Y-Yorba, J.*: Talk at the Daresbury Study Weekend (June 1970), Daresbury Report No. DNPL/R 7 (1970).
43. *SLAC-Berkeley-Tufts Collaboration.* Phys. Rev. Letters **23**, 498 (1969).
44. *Meyer, H., et al.*: Phys. Letters **33** B, 189 (1970).
45. *Caldwell, D. O., et al.*: Phys. Rev. Letters **25**, 609 (1970).
46. See e.g.: *Damashek, M.*, *Gilman, F. G.*: SLAC Report No. SLAC-PUB-697 (1969); – *Meyer, H., et al.*: DESY Report No. DESY 70/17 (1970).
47. *Buschhorn, G., et al.*: Phys. Letters **33** B, 241 (1970); – *Anderson, R. L., et al.*: Phys. Rev. Letters **25** B, 1218 (1970).
48. See e.g.: *Manweiler, R.*, *Schmidt, W.*: Phys. Letters **33** B, 366 (1970).
49. *Massam, T.*, *Zichichi, A.*: Nuovo Cimento **43**, 1137 (1966).
50. For a discussion of the experimental situation see e.g.: *Silverman, A.*: op. cit. in Ref. [34].
51. *Di Vecchia, P.*, *Drago, F.*: Lett. Nuovo Cimento **1**, 917 (1969).
52. *Cordes, J. G.*, *O'Donnell, P. J.*: Univerity of Toronto preprint (1969).

Dr. *Günter Wolf*
DESY
2000 Hamburg, Germany

Meson Photoproduction on Nuclei

L. Foà

Contents

In the last years the coherent photoproduction on complex nuclei at energies of a few GeV and above, has proved to be an important tool in the study of the properties of bosons and on their interactions with matter. This reaction shows important differences with respect to the photoproduction on single nucleons. The photoproduction cross-section for a boson B on a nucleon can be written as the sum of a term independent from the spin and the isospin of the nucleon, $\left(\frac{d\sigma}{dt}\right)_N^{(0)}$, and a spin and isospin dependent term, $\left(\frac{d\sigma}{dt}\right)_N^{(1)}$:

$$\left(\frac{d\sigma}{dt}\right)_N = \left(\frac{d\sigma}{dt}\right)^{(0)} + \left(\frac{d\sigma}{dt}\right)^{(1)}$$

Then the cross-section for coherent photoproduction of the same boson on a nucleus of atomic number A and of spin 0 can be expressed as:

$$\left(\frac{d\sigma}{dt}\right)_A = N_{\text{eff}}^2(A, \sigma_{BN}, \beta_B, t)\left(\frac{d\sigma}{dt}\right)_N^{(0)} \tag{1}$$

where $-t$ is the invariant four-momentum transfer squared and N_{eff} is the form factor of the nucleus, normalized to A nucleons, written in such a way as to take into account the reabsorption of the produced boson along its way off the nucleus; σ_{BN} is the total cross-section of the boson on nucleons, and β_B is the ratio between the real and the imaginary parts of the forward boson nucleon scattering amplitude.

The interesting points in expression 1) are the following:

1. The coherence acts as a filter to separate the spin and isospin independent term in the production amplitude on nucleons [1].

2. If $\left(\frac{\mathrm{d}\sigma}{\mathrm{d}t}\right)_N^{(0)}$ is the dominant term in $\left(\frac{\mathrm{d}\sigma}{\mathrm{d}t}\right)_N$, and in heavy nuclei such that $N_{\mathrm{eff}}^2 \gg 1$, one has

$$\left(\frac{\mathrm{d}\sigma}{\mathrm{d}t}\right)_A (t\sim 0) \gg \left(\frac{\mathrm{d}\sigma}{\mathrm{d}t}\right)_N (t\sim 0).$$

Many experiments took advantage of this enhancement of the cross-section to study rare production modes like the measurements of the π^0 and η lifetimes through Primakoff effect [2–4] or rare branching ratios (typical examples have been the experiments on the leptonic decay rates of vector bosons [5]).

3. A study of the A dependence of N_{eff}^2 allows the determination of the total cross-section on nucleons for short living bosons.

The last property has been extensively applied to determine the total cross-sections of vector mesons on nucleons. We shall begin the present review with a discussion of this important point.

The coherent amplitude for vector meson photoproduction on a nucleus can be written, in an optical model approach, as [6]

$$f_c = f_0 \cdot 2\pi \int_0^\infty b\,\mathrm{d}b \int_{-\infty}^{\infty} \mathrm{d}z\,\varrho(b,z)\,J_0(b,q_\perp)\,\mathrm{e}^{iq_\parallel z} \exp\left\{-\frac{\sigma_{VN}}{2}(1-i\beta_V)\int_z^\infty \varrho(b,z')\,\mathrm{d}z'\right\} \tag{2}$$

where f_0 is the spin and isospin independent part of the photoproduction amplitude of the boson V on nucleons, b and z are coordinates respectively orthogonal and parallel to the direction of flight of the incoming photon, $\varrho(b,z)$ is the nuclear matter distribution, $q_\parallel$ and $q_\perp$ are the parallel and transversal components of the momentum transfer to the nucleus. $q_\parallel$ gives the minimum momentum transfer which is needed to photoproduce the mass of the vector meson in the forward direction ($q_\parallel = m_V^2/2E_V$). The second exponential describes the effects of the final state interaction of the boson in the nucleus. The exponent contains two terms: the first term is real and accounts for the attenuation that each wave produced on a single nucleon feels in its way off the nucleus. The second part of the exponent is a phase term containing the ratio between the real and the imaginary parts of the forward boson nucleon scattering amplitude β_V. Its effect can be understood by remembering that coherence requires all single waves produced on different nucleons to be in phase when leaving the nucleus. A rotation of these phases can increase or reduce the degree of coherence according to the sign of β_V. The effect of this term depends on $\int_z^\infty \varrho(b,z')\,\mathrm{d}z'$, which in turn depends on A, the mass number of the nucleus. Therefore, if

$\beta_V \neq 0$, a modification of the A-behaviour of the photoproduction cross-section will be produced, similar to that introduced by the first term of the exponent.

Aim of all the experiments we are going to discuss is the determination of σ_{VN} by an analysis of the photoproduction data in terms of expression (2). Therefore all other parameters entering this expression must be independently measured. These parameters are the nuclear matter distribution $\varrho(A)$ and β_V.

ϱ^0 Photoproduction

At the present time, the meson for which we have the largest amount of informations is the ϱ^0. Between 1967 and 1969 three experiments were performed by *Asbury et al.* at DESY [7], *McClellan et al.* at Cornell [8], and by *Bulos et al.* at SLAC [9]. The results were largely inconsistent essentially because the procedures adopted by the groups to analyse the data were different and because the fundamental role played by β_V was not recognized. This year three "second generation" experiments

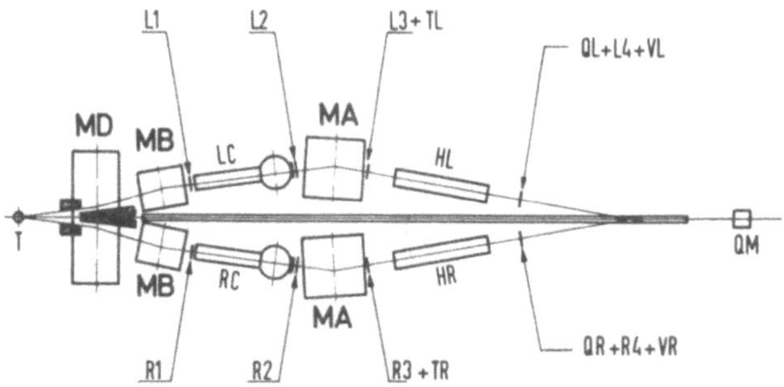

Fig. 1. General lay-out of the DESY-MIT experiment

have been performed by *Alvensleben et al.* (DESY-MIT) [10], *McCellan et al.* (Cornell) [11], *Behrend et al.* (Rochester) [12], and the data by *Bulos et al.* have been reanalyzed. Each group has measured directly the nuclear radii. The DESY-MIT group has produced the most accurate results by studying the ϱ^0-photoproduction on 13 nuclei from 4.8 to 7.2 GeV with a very high statistical precision ($\sim 10^6$ events in total). Fig. 1 shows the experimental apparatus used to detect the $\pi^+\pi^-$ pairs.

A two-arm spectrometer measures direction, momentum and mass of the two emitted particles, using threshold Cerenkoff counters and scintillation counter hodoscopes. Since the spectrometer recombines particles with the same $p\theta$, a large acceptance in momentum and mass of the pion pairs is obtained, together with a good resolution in these variables, as shown in Table 1.

Table 1. *Typical acceptance and resolutions in the DESY-MIT experiment*

Acceptance	Resolution
$\frac{\Delta m}{m} = \pm .10$	$\delta m = \pm 15$ MeV
$\frac{\Delta p}{p_0} = \pm .18$	$\delta p = \pm 150$ MeV/c
$\frac{\Delta \theta}{\theta} = \pm .14$	$\delta t_1 = \pm .001\,(\mathrm{GeV}/c)^2$

In Fig. 2 a sample (2%) of the data is plotted for all 13 nuclei against $m(\pi^+\pi^-)$ and $t_\perp$. The mass spectra show the ϱ^0 bump superimposed to a background which decreases with increasing A. The t-distributions show a clear peak in the forward direction which shrinks from Beryllium to Uranium followed by a flatter region corresponding to incoherent ϱ^0-production. It is clear that these data can allow a determination of the nuclear radii only if they are analysed in terms of expression (2) in which other parameters like $\sigma_{\varrho N}$, β_ϱ are present and for which therefore reasonable assumptions have to be made. In this experiment $\sigma_{\varrho N}$ was set equal to 26.7 mb and β_ϱ equal to -0.2. Furthermore the true ϱ^0's have to be separated from the background of non-resonating $\pi^+\pi^-$ pairs contained in the mass spectra.

The analysis was performed by keeping only the central region of the mass distributions ($690 < m < 860$ MeV/c^2) and only the data at $t < 0.01\,(\mathrm{GeV}/c)^2$ assuming that in these regions of m and t the pion pairs from ϱ^0 decay dominate on background. For safety the mass interval was splitted into 6 bits and the analysis was repeated for each of them. The nuclear density distribution was assumed to be described by the Wood Saxson formula

$$\varrho = \varrho_0 \left[1 + \exp\frac{r - R(A)}{s}\right]^{-1}$$

with a fixed value of $s = 0.545$ fermi.

A comparison of the results showed that for $A > 27$ the radii were consistent for all mass bits and did not indicate any trend. Furthermore

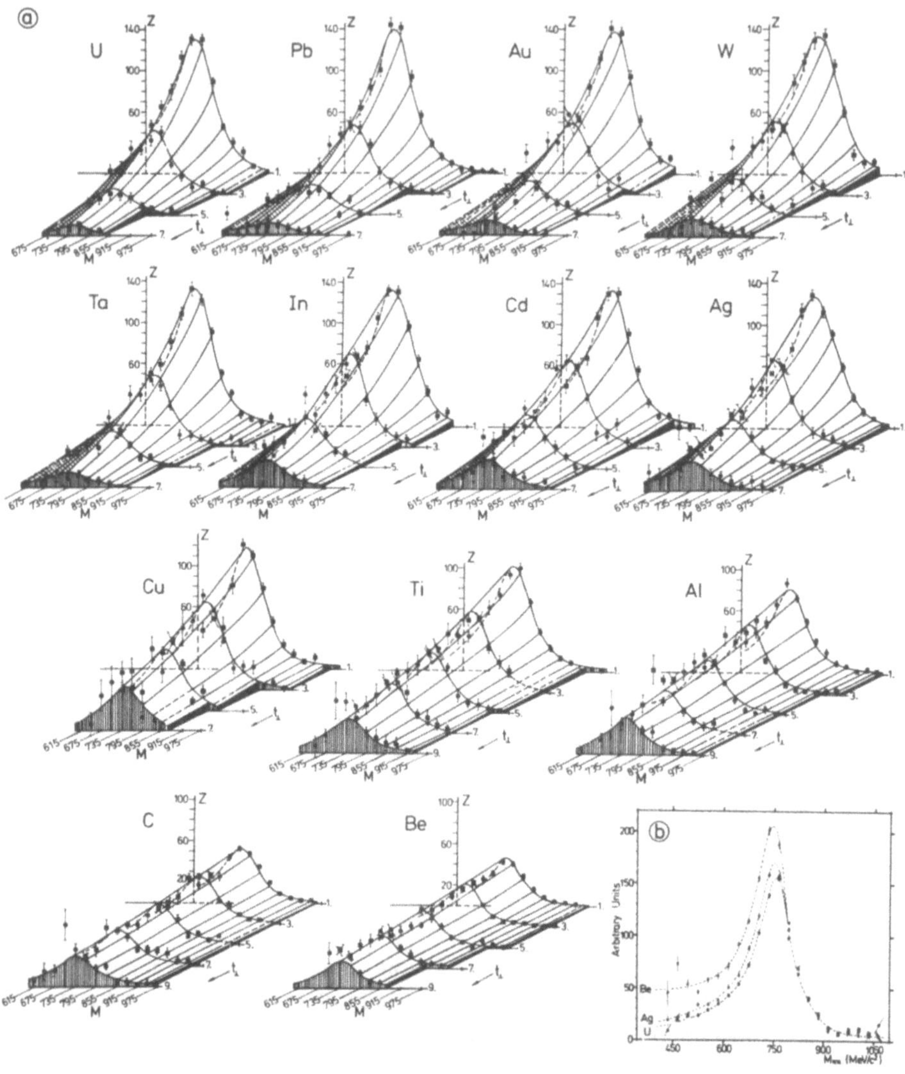

Fig. 2. Three dimensional plots of the mass and t distributions obtained in the DESY-MIT experiment

large variations of the parameters in formula (2) could effect the radii less than 2%. For $A < 27$ the background affected more the results and a larger error was attributed to the measured radii. As a general remark we can say that the radii determined in all quoted experiments are in good agreement for $A > 27$. On the other hand, for light nuclei each group uses different density distributions and different radii. In Fig. 3 the results of the DESY-MIT measurements are shown together with a fit obtained using the expression $R(A)=r_0 A^{1/3}$.

The best value is $r_0 = 1.12 \pm 0.02$ fermi. This is larger than, but not inconsistent with $r_0 = 1.07$ fermi as determined in electron scattering experiments [13].

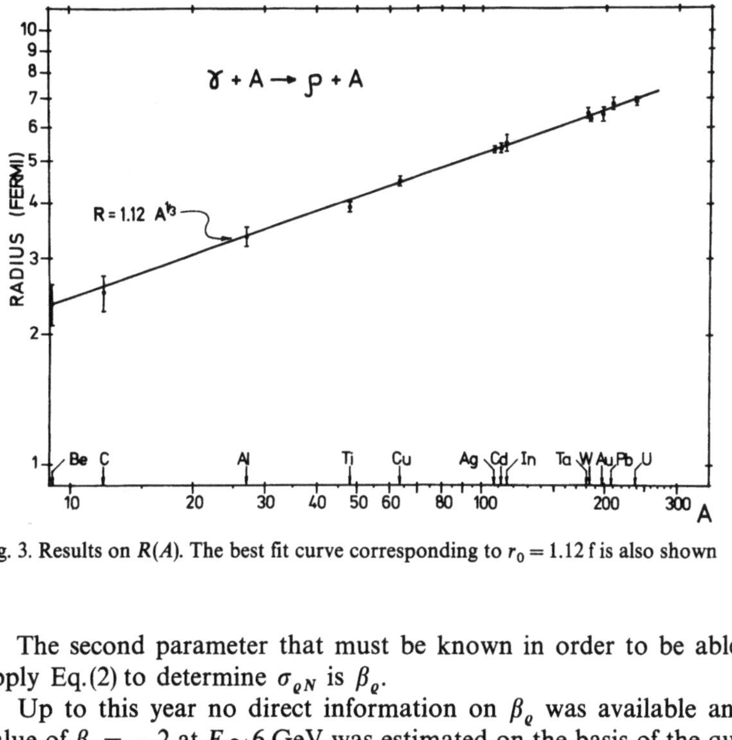

Fig. 3. Results on $R(A)$. The best fit curve corresponding to $r_0 = 1.12$ f is also shown

The second parameter that must be known in order to be able to apply Eq. (2) to determine $\sigma_{\varrho N}$ is β_ϱ.

Up to this year no direct information on β_ϱ was available and a value of $\beta_\varrho = -.2$ at $E \sim 6$ GeV was estimated on the basis of the quark model prediction

$$\beta_{\varrho^0} = \beta_{\pi^0} = \beta_\gamma .$$

Very recently a measurement of this quantity has been performed at DESY by the same group, by studying the photoproduction of electron-positron pairs on Berillium at $E_\gamma \sim 6$ GeV [14]. The main contributions to this channel come from wide angle electron pair production (Bethe-Heitler) and from ϱ^0 photoproduction followed by $\varrho^0 \to e^+ e^-$ decay. The interference between these two terms depends on the phase of the ϱ^0 photoproduction amplitude, being the Bethe-Heitler amplitude real. This interference term is antisymmetric with respect to the exchange of the electron and positron momenta p_+ and p_-. Therefore its presence would show up in an asymmetry in the distribution of events when plotted as a function of a variable antisymmetric in p_+ and p_-, as for instance $p_+\theta_+ - p_-\theta_- = \delta$ (δ represents the transverse momentum of the pair). Unfortunately the expression of the overall cross-section is extremely sensitive to the shape of the ϱ^0 mass distribution and to the parameters entering the ϱ^0 photoproduction cross-section ($m_\varrho, \Gamma_\varrho, \ldots$). In order to minimize the dependence on these parameters, a comparison was performed between the following quantities:

$$\frac{\sigma_i(m,\delta)}{\sqrt{\sigma_{\mathrm{BH}}(m)\cdot\sigma_{\mathrm{C}}(m)}} \quad \text{(calculated)}$$

and

$$\frac{N_+(m,\delta)-N_-(m,\delta)}{\sqrt{N_{\mathrm{BH}}(m)\cdot N_{\mathrm{C}}(m)}} \quad \text{(measured)}$$

where $\sigma_i(m,\delta)$ is the cross-section for the interference term, N^+ and N^- are the rates obtained by exchanging positron and electron at the same value of δ, σ_{BH} and N_{BH} are respectively the cross-section and the rate for Bethe-Heitler pair production and σ_{C} and N_{C} are the cross-section and the rate for ϱ^0 photoproduction in the e^+e^- channel. The kinematical conditions were such that σ_{C} and σ_{BH} were nearly equal. In Fig. 4 the difference $N_+ - N_-$ is plotted against δ, for various mass intervals, and shows clearly the presence of an asymmetric term.

Once derived from the above results the phase for ϱ^0 photoproduction on Beryllium, the Margolis theory [15] allows to calculate the phase of the ϱ^0-nucleon scattering amplitude. The final result is $\varphi = 11.8 \pm 4.4$ which corresponds to a value $\beta_\varrho = -0.2 \pm 0.1$ in good agreement with the quark model prediction.

In this meeting, the results of a similar experiment performed at Nina at a somewhat lower energy have been announced. The value obtained for the ϱ^0 phase is $\varphi = 19^{\circ}{}^{+8.5}_{-7.5}$, not in contradiction with the DESY-MIT result.

Once $R(A)$ and β_ϱ are known, the last problem consists in the separation of the diffractively produced ϱ^0's from backgrounds in the mass and t-distributions. I shall describe here the procedures followed by the Cornell and by the DESY-MIT groups. In the first experiment a definite model has been adopted to interpret the shape of the pion-pair mass distributions. Two processes have been assumed to be dominant: the ϱ^0 photoproduction, described by a Breit-Wigner formula $BW(m)$ and a non resonant pion pairs production, $D(m)$.

The interference between these two amplitudes $I(m)$ (Söding term) is responsible for the asymmetry in the ϱ^0 mass spectrum. The absolute value of the cross-section for the non resonant production was calcu-

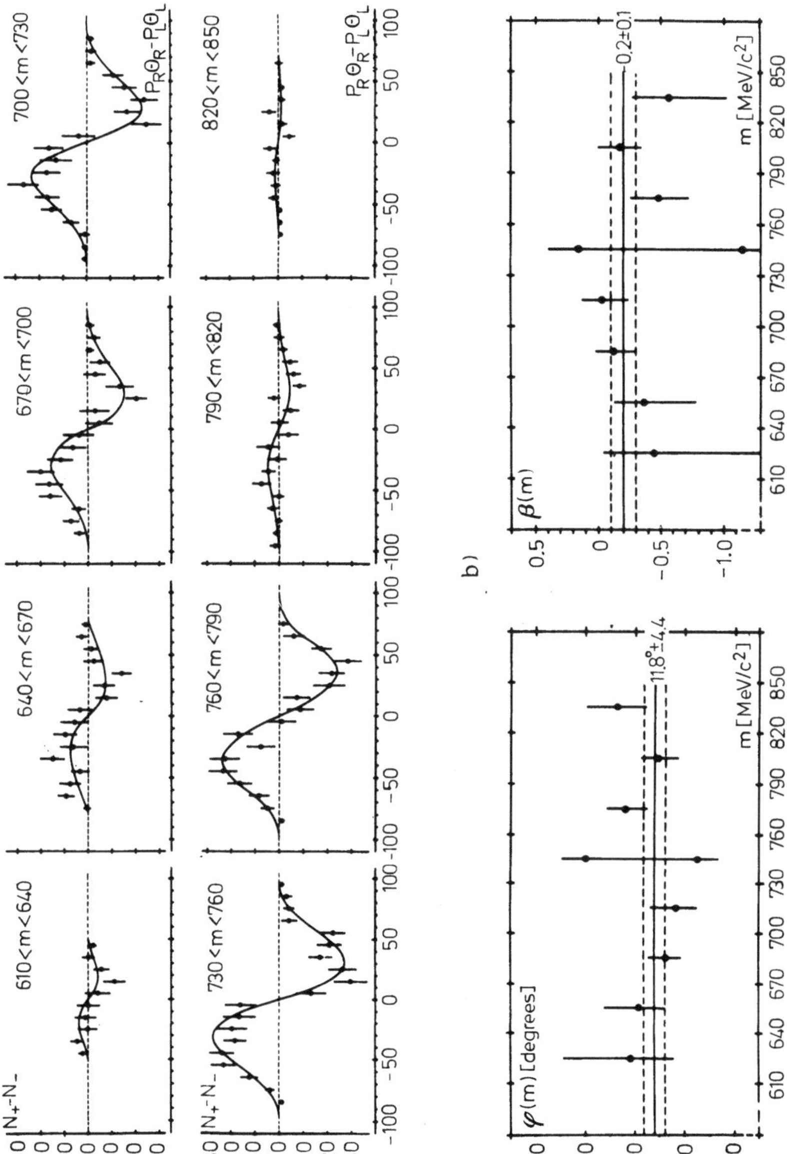

Fig. 4. Difference $(N_+ - N_-)$ as function of δ and $m_{\pi^+\pi^-}$. Fig. b) shows the results obtained on φ and β for each mass bin

lated and a correction was applied to avoid double counting [16]. A factor C_0 was left free in front of the interference term and determined in the best fit analysis. Only for Deuterium an incoherent background was subtracted in the t-distributions. For heavier nuclei this contribution

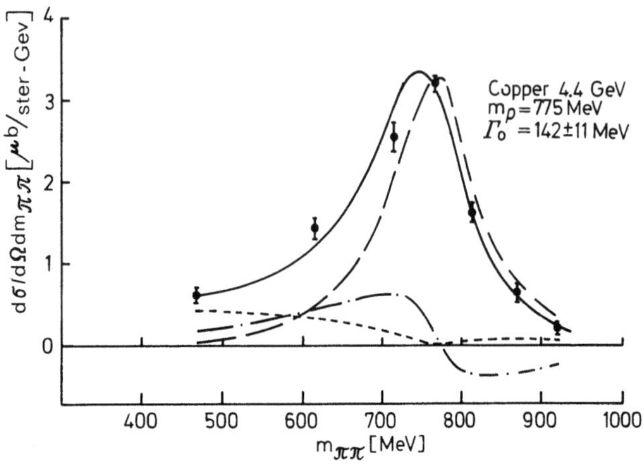

Fig. 5. Mass spectrum for copper measured in the Cornell experiment. Full line shows the best fit to the data obtained by adding ϱ^0 photoproduction (——), non resonant $\pi^+\pi^-$ production (- - -) and Söding term (- · - · - · -)

was experimentally found to be negligible. Fig. 5 shows the measured mass spectrum and the separate contributions of the three terms.

In the analysis of the DESY-MIT results the experimental distributions in m, $t_\perp$ and p have been fitted with the following expression:

$$\frac{\mathrm{d}\sigma}{\mathrm{d}\Omega\,\mathrm{d}m}(A,p,t,m) = \frac{1}{\pi}\,p^2\,2m\,R_N(m)\,(f_c^2 + f_{\mathrm{inc}}^2) + B\,G(A,p,t,m)$$

where f_c is the coherent photoproduction amplitude described by formula 2), f_{inc} is the incoherent photoproduction amplitude, as calculated by *Trefil* [21], and the background function $B\,G(A,p,m,t)$ is the product of polynomials in m, t and p. $R_N(m)$ represents the various formulas used to describe the $\pi^+\pi^-$ mass distribution. Five models have been tested the most interesting of which are:

$$\text{I)}\quad R_1(m) = \left(\frac{m_0}{m}\right)^n \cdot B\,W(m) \text{ with } n=4 \quad \text{(Ross-Stodolsky mechanism)}$$

$$\text{II)}\quad R_2(m) = B\,W(m) + I(m) \qquad \text{(Söding mechanism)}$$

$$\text{III)}\quad R_3(m) = R_1(m) + I(m)$$

where $B\,W(m)$ and $I(m)$ have the meaning already outlined in the description of the Cornell experiment. These expressions need some comments. Recently the ϱ^0 photoproduction cross-section on Hydrogen has been measured in the SLAC Bubble chamber by *Bingham et al.* [17] using a polarized photon beam. The mass spectra of the pion pairs have been analysed according to the previous models and in Eq. I) the exponent n was allowed to vary as a function of t. The result, plotted in Fig. 6, shows that n depends strongly on t. This dependence forbids the use of the Ross-Stodolsky mass distribution in the analysis of ϱ^0-photoproduction data obtained on Hydrogen; on the other hand, formula I) can be applied to analyze data on heavier nuclei, for which the form factor cuts the t distribution at $t<0.01\ (\mathrm{GeV}/c)^2$.

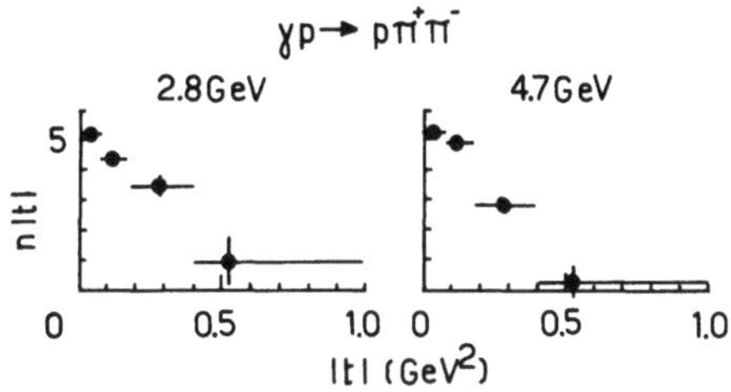

Fig. 6. Dependence of the exponent $n(t)$ on t as determined in the Bingham et al. experiment

The best-fit analysis of the data mentioned above showed that the three models are equally favoured and that the final results of the experiment, calculated according to the three models, were consistent within the quoted errors. The background function $BG(A, p, t, m)$ was found to be a second order polynomial in m, and to be insensitive to the other variables.

Once the true ϱ^0's were separated from background, the analysis could proceed to the determination of $|f_0|^2$ and of $\sigma_{\varrho N}$. In all experiments the value of the cross-section at $\theta_\varrho = 0$ was unfolded by making use of the known values of the nuclear radii and of the resolution of the apparatus. This cross-section was then plotted as functions of A and fit with formula (2), by optimising the values of $|f_0|^2$ and of $\sigma_{\varrho N}$. Fig. 7 shows the results for the DESY-MIT and for the Cornell experiments. The plots show very clearly the effect of the ϱ^0 reabsorption which suppresses the production for large values of A.

The results for $|f_0|^2$ and for $\sigma_{\varrho N}$ obtained by the four experiments [9–12] are given in Table 2. The agreement among the various results is rather good and supports the quark model prediction

$$\sigma_{\varrho N} = \sigma_{\pi N} = 27\ \mathrm{mb}.$$

In the framework of the vector dominance model, the value of $|f_0|^2$ can be used to calculate the ϱ-photon coupling constant $\gamma_\varrho^2/4\pi$, through the relation

$$|f_0|^2 = \frac{\alpha}{4}\left(\frac{\gamma_\varrho^2}{4\pi}\right)^{-1} \frac{1}{16\pi}\,\sigma_{\varrho N}^2(1+\beta_\varrho^2).$$

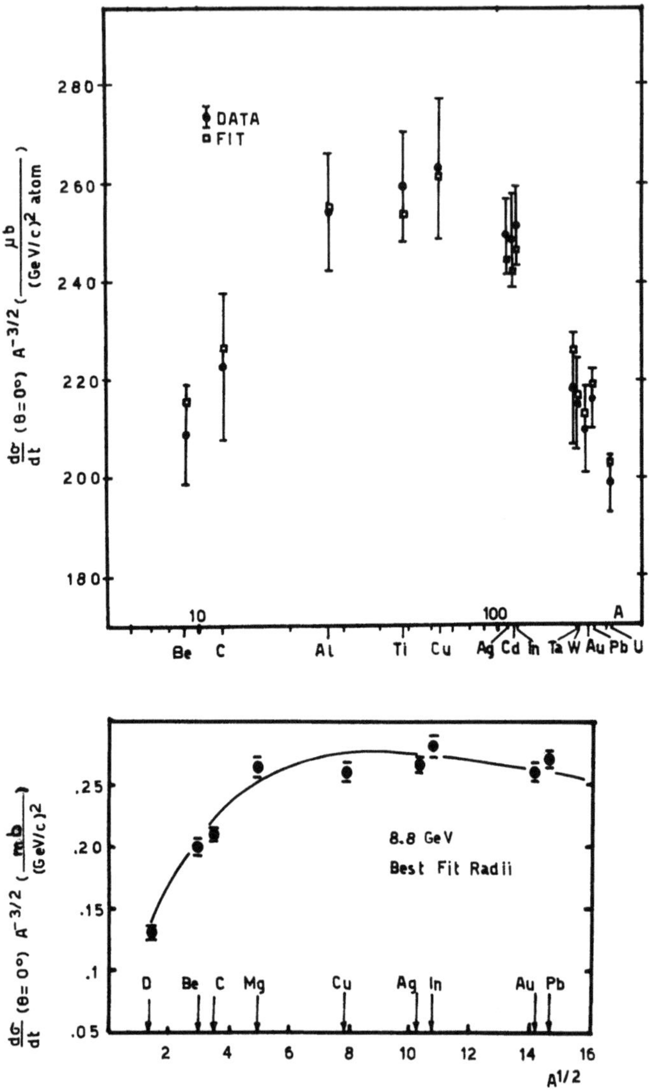

Fig. 7. $A^{-3/2} \cdot d\sigma/dt(\theta=0)$ plotted versus A for the DESY-MIT experiment (upper graph) and for the Cornell experiment (lower graph)

It is clear that the determination of $\gamma_\varrho^2/4\pi$ depends critically on the absolute value of the cross-section, in contrast to the measurement of $\sigma_{\varrho N}$ which demands only the knowledge of the relative A-dependence of the cross-section. The last column of Table 2 gives a comparison of the results on $\gamma_\varrho^2/4\pi$, as obtained by the various groups.

Table 2. *Compilation of results on ϱ^0 interaction parameters*

Experiment	E(GeV)	$\lvert f_0 \rvert^2$ (μb/GeV/c)^2	$\sigma_{\varrho N}$(mb)	$\frac{\gamma_\varrho^2}{4\pi}$
DESY-MIT	4.8→7.2	118±6	26.7±2	0.57±0.1
Cornell	6.1	117±10	26.1±.9	0.58±0.03
	6.5	124±15	30.1±1.5	0.74±0.05
	8.8	105±11	26.8±1.2	0.68±0.04
Rochester	8.0	117±8	26.8±2.4	0.62±0.1
SLAC	6–12–18		28.8±2.0	0.85±0.1
$\varrho \leftrightarrow e^+e^-$	—	—	—	0.48±0.04

ω-Photoproduction

Only this year the results of the first two experiments on ω photoproduction have been published. The reason for this delay with respect to ϱ^0 photoproduction experiments has to be attributed mainly to the difficulty of detecting the dominant decay channels of the ω mesons, $\pi^+\pi^-\pi^0$ and $\pi^0\gamma$. The first experiment has been performed at Cornell by *Behrend et al.* [18] (Rochester group), with the same apparatus used to study ϱ^0 photoproduction; ω mesons of about 6.8 GeV were produced on Be, C, Cu and Pb and detected through $\pi^+\pi^-\pi^0$ decay. The first part of the set up (shown in Fig. 8) consisted of a large magnet and of a system

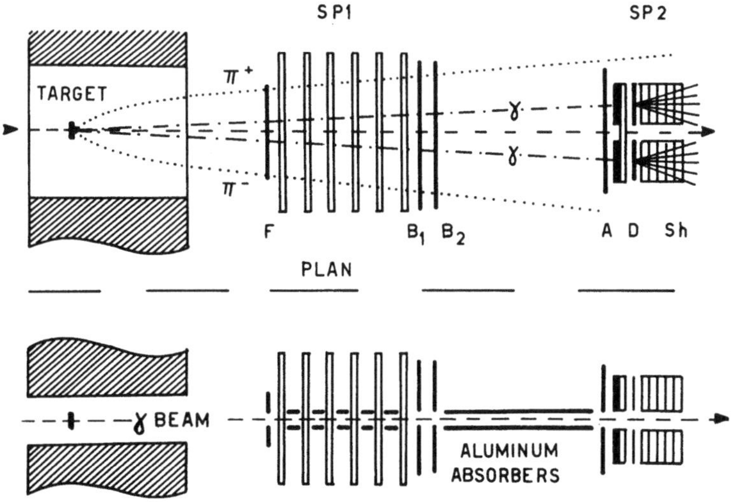

Fig. 8. General lay-out of the Behrend et al. experiment on ω photoproduction. The same set-up, without photon detectors was used to study ϱ^0 photoproduction

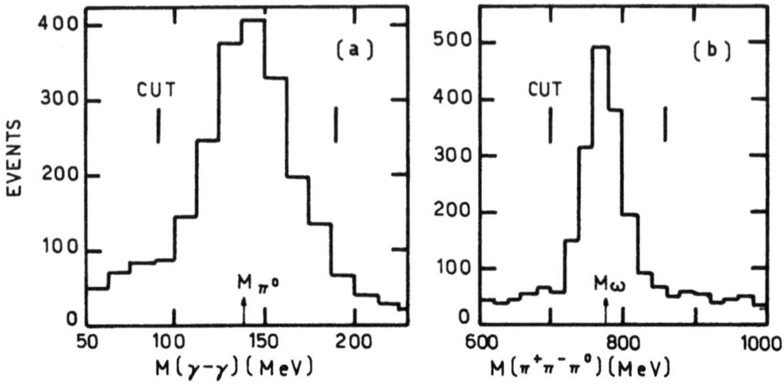

Fig. 9. $\gamma\gamma$ and $\pi^+\pi^-\pi^0$ mass spectra showing the π^0 and ω peaks

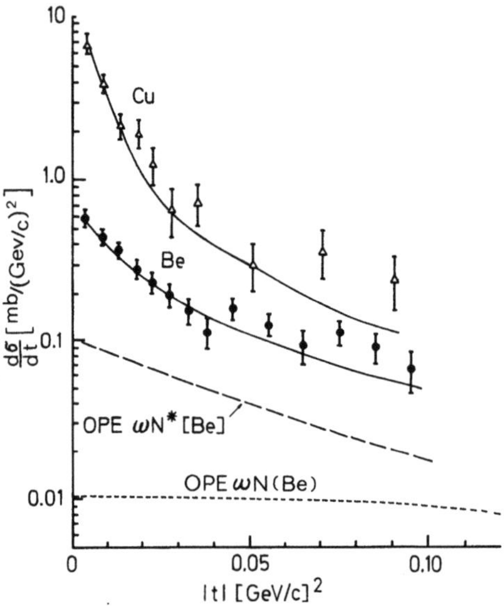

Fig. 10. Angular distributions for beryllium and copper. The relative contributions of ωN^* and ωN are also shown for beryllium

of spark chambers and counters measuring direction and momentum of the charged pions. Photons from π^0 decay were also detected, after conversion, and their energy was measured by shower counters. π^0's were clearly separated from the background in the $\gamma\gamma$ mass spectra (Fig. 9) and their momenta, together with the measured π^+ and π^- momenta, allowed a good reconstruction of the ω mass. Fig. 10 shows the angular distributions measured for beryllium and copper. A coherent peak and an incoherent background can clearly be seen. The subtraction of this background is a difficult point for all ω-photoproduction experiments. It is already known from Hydrogen bubble chamber measurements that a large fraction of ω-photoproduction on nucleons proceeds through unnatural parity exchange. This part of the amplitude does not contribute to the coherent cross-section, while contributing to the incoherent ω production. Furthermore other processes like $\gamma+N\to\omega+N^*$ or $\gamma+N\to\varrho^0+N$ followed by $\varrho^0\to\omega$ rescattering can increase appreciably the incoherent cross-section. Therefore the incoherent background to be subtracted under the coherent peak is 2–3 times larger then in the ϱ^0-photoproduction experiments. The difficulty of this subtraction consists in the different t behaviour of the process $\gamma+N\to\omega+N$ and $\gamma+N\to\omega+N^*$ in nuclear matter, the first one being suppressed in the forward direction by a form factor taking into account the Pauli exclusion principle.

The Rochester group calculates the relative contributions of the two terms according to an OPE model [19]; the term with excitation of N^* is found to be dominant.

After subtraction of the incoherent production, data were fitted with expression (2), using for the nuclear radii the values determined in the study of ϱ^0-photoproduction and setting $\beta_\omega=-.2$. The best fit analysis gave the following results

$$\begin{cases}\sigma_{\omega N}=33\pm5.5\text{ mb}\\ |f_0|^2=(11\pm1.9)\,\mu\text{b/GeV}^2\end{cases}$$

and, through vector dominance,

$$\frac{\gamma_\omega^2}{4\pi}=9.5\pm2.1\,.$$

Being the value of $\sigma_{\omega N}$ compatible with the quark model prediction $\sigma_{\omega N}=\sigma_{\varrho N}=27$ mb, the analysis was repreated using $\sigma_{\omega N}=27$ mb as an extra constraint and the best fit values were found to be

$$\begin{cases}|f_0|^2=(9.6\pm1.2)\,\mu\text{b/GeV}^2\\ \dfrac{\gamma_\omega^2}{4\pi}=7.3\pm1.0\,.\end{cases}$$

The second experiment performed at DESY by *Braccini et al.* (Bonn-Pisa Group [20]) follows a different approach. Angular distributions for ω photoproduction have been measured at $\sim$5.7 GeV on C, Al, Zn, Ag, Ta and Pb. The detection apparatus, shown in Fig. 11, consists of a set of total absorption Cerenkoff counters searching for two of the three photons of the $\omega\to\pi^0\gamma$ decay. Fig. 12 shows a correlation plot between the energies of the two detected photons. The plot on the right

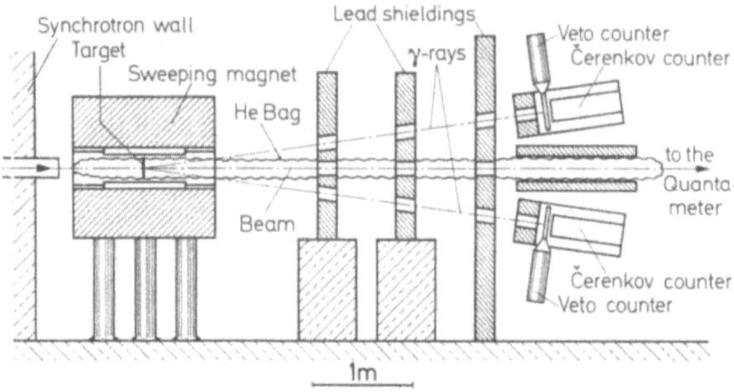

Fig. 11. Experimental set-up used by the Bonn-Pisa group

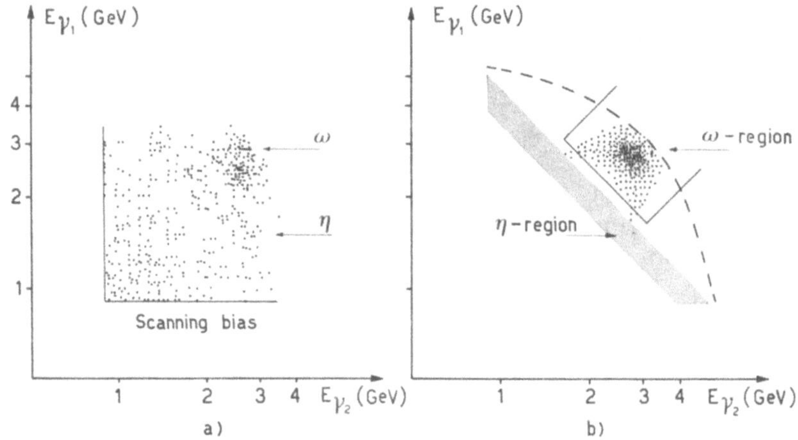

Fig. 12. Calculated and measured $E_{\gamma 1}/E_{\gamma 2}$ correlation plots

hand side describes the expected distribution of events, as calculated with the Montecarlo method. On the left hand side the experimental plot shows a clear cluster of points in the ω region, separated from the $\eta \to \gamma\gamma$ events and from the low energy background. The angular distributions obtained in the experiments are plotted in Fig. 13. Also in these graphs, the coherent peak and the incoherent background can clearly be distinguished. Full lines represent the best fit curves obtained by folding the cross-section described by expression (2) in the detection efficiency.

Due to the absence of an unique theory of the t-dependence of the incoherent ω photoproduction under the forward peak, the analysis was repeated calculating the incoherent cross-section according to four different models [21]. In this way the sensitivity of the results on the uncertainty in the background subtraction could be tested. Furthermore,

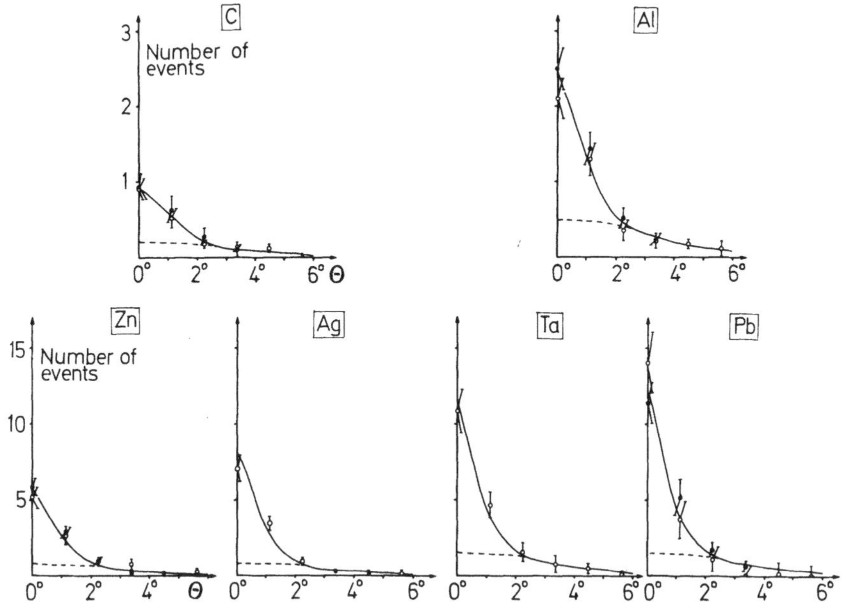

Fig. 13. Angular distributions for ω yield in various elements, expressed in units of counts/[(nucleus/cm^2) $\times$ 10^{35} eq. q.]

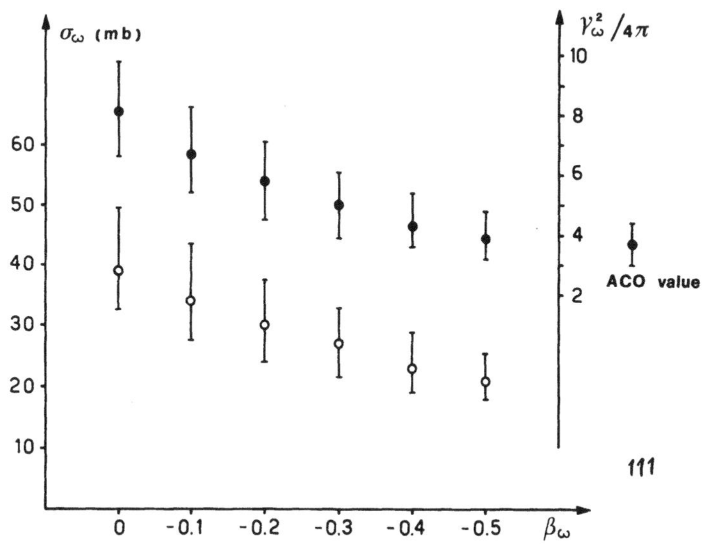

Fig. 14. Dependence of $\sigma_{\omega N}$ (open points) and $\gamma_\omega^2/4\pi$ (full points) on β_ω. All these values are equally favoured by the χ^2 analysis

being the real part of the ω-nucleon forward scattering amplitude still unknown, the analysis was repeated for various values of β_ω. In Fig. 14 the best fit values of $\sigma_{\omega N}$ and $\gamma_\omega^2/4\pi$ are plotted against β_ω, the t-dependence of the incoherent cross-section being calculated according to the results obtained by *Bojarski et al.* [21] for π^+ and π^- photoproduction on nuclei. For $\beta_\omega = -.2$ the results of the analysis of the data in terms of expression (2) are

$$\begin{cases} \sigma_{\omega N} = (30^{+7}_{-6})\ \text{mb} \\ |f_0|^2 = (15{,}1 \pm 3{,}1)\ \mu\text{b/GeV}^2 \\ \dfrac{\gamma_\omega^2}{4\pi} = 5{,}8 \pm 1{,}3\,. \end{cases}$$

A different choice of the t-dependence of the incoherent cross-section under the coherent peak can change $\sigma_{\omega N}$ by ± 5 mb and increase $\gamma_\omega^2/4\pi$ up to 6.8. An other source of uncertainty on the value of $\gamma_\omega^2/4\pi$ is the possible contribution of $\varrho^0 \to \pi^0\gamma$ to the $\pi^0\gamma$ yield. With the present upper limit [22] $\dfrac{\Gamma\varrho^0 \to \pi^0\gamma}{\Gamma\varrho^0 \to \pi^+\pi^-} < .2\%$ only the interference term between ω and ϱ^0 photoproduction is relevant and its importance depends on the relative phase φ of the ω and ϱ^0 photoproduction amplitudes, ranging from 0 (for $\varphi = 90°$) to $\sim 30\%$ (for $\varphi = 0°$).

Table 3 gives the values obtained up to now on $\sigma_{\omega N}$ and $\gamma_\omega^2/4\pi$. The values of $\sigma_{\omega N}$, also is affected by large uncertainties, are consistent with the quark model predictions. The values of $\gamma_\omega^2/4\pi$ appear larger than the result obtained at ACO with the electron positron storage rings, $\gamma_\omega^2/4\pi = 3.7 \pm 0.7$ [23]. A clear answer to this question will come only from the future "second generation" experiments.

Table 3. *Compilation of results on ω interaction parameters. Also the results obtained by Ballam et al.* [25] *in the SLAC Hydrogen Bubble Chamber with polarized photons are given for comparison*

Experiment	Energy (GeV)	$\lvert f_0\rvert^2 (\mu\text{b/GeV}/c)^2$	$\sigma_{\omega N}$(mb)	$\gamma_\omega^2/4\pi$	Remarks
Behrend et al.	6.8	11.4 ± 1.9	33.5 ± 5.5	9.5 ± 2.1	$\beta = -0.2$
$\omega \to 3\pi$		9.6 ± 1.2	—	7.3 ± 1.0	$\sigma_{\omega N} = 27$ mb
Braccini et al.	5.7	15.1 ± 3.1	30^{+7}_{-5}	5.8 ± 1.3	$\beta = -0.2$
$\omega \to \pi^0\gamma$			(26–35)		(no $\varrho \to \pi^0\gamma$)
Ballam et al.	4.8	15.2 ± 3.8	—	4.6 ± 1.4	$\sigma_\omega = 27$ mb
Augustin et al.	—	—	—	3.7 ± 0.7	—

Primakoff Effect

The measurement of the 2γ decay width of pseudoscalar bosons through their electromagnetic photoproduction on nuclei (Primakoff effect) [24] is a nice example of the help that nuclei can provide to carry into evidence small cross-sections. The basic idea of the Primakoff effect consists in the fact that the diagram for boson photoproduction in an electric field, shown in Fig. 15a), contains the same vertex function appearing in the decay diagram (Fig. 15b), the only difference lying in the non zero value

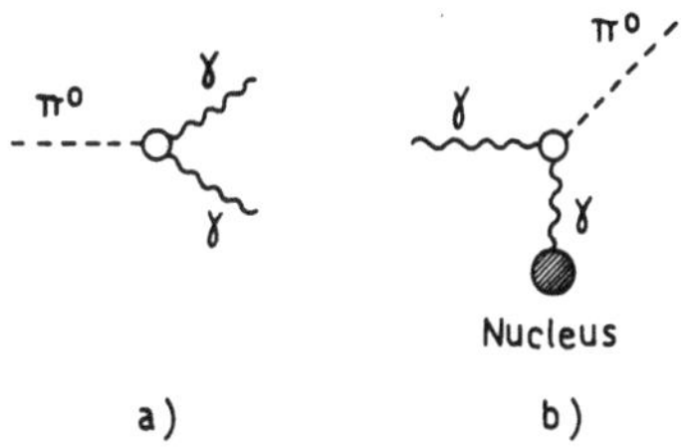

Fig. 15. Diagrams for π^0 decay and for Primakoff effect

of the mass of the virtual photon exchanged with the field source. At high energy and for rather low masses of the bosons (π^0 and η) the virtual mass is very small (always less than 100 MeV) and any appreciable variation of the coupling constant over such a small range is very unlikely to be expected. The difficulty of the experiment consists in the separation of the electromagnetic term from the nuclear photoproduction cross-section, because the two processes interfere with an unknown phase. The use of nuclei as targets provides again a powerful tool to discriminate the two processes. The coherence of the process increases by a factor Z^2 the cross section for electromagnetic production, which is very small on single nucleons, without introducing any reabsorption on the outgoing bosons, since the production takes place outside the nucleus. Furthermore it has been already pointed out that the requirement of coherence acts a selection on the nuclear photoproduction amplitudes, cancelling all terms flipping the spin or the isospin of a nu-

Table 4. *Comparison of the different dependences of electromagnetic and nuclear productions on E_{π^0}, θ_{π^0} and on the atomic number of the target nucleus*

Variable	Primakoff effect	Nuclear production
θ_{π^0}	peak at $\sim .1°$	peak at 2°–4°
E_{π^0}	$E_{\pi^0}^4$ (at the peak)	$\sim$ constant
Atomic number	Z^2	$\sim A$ (at the peak)

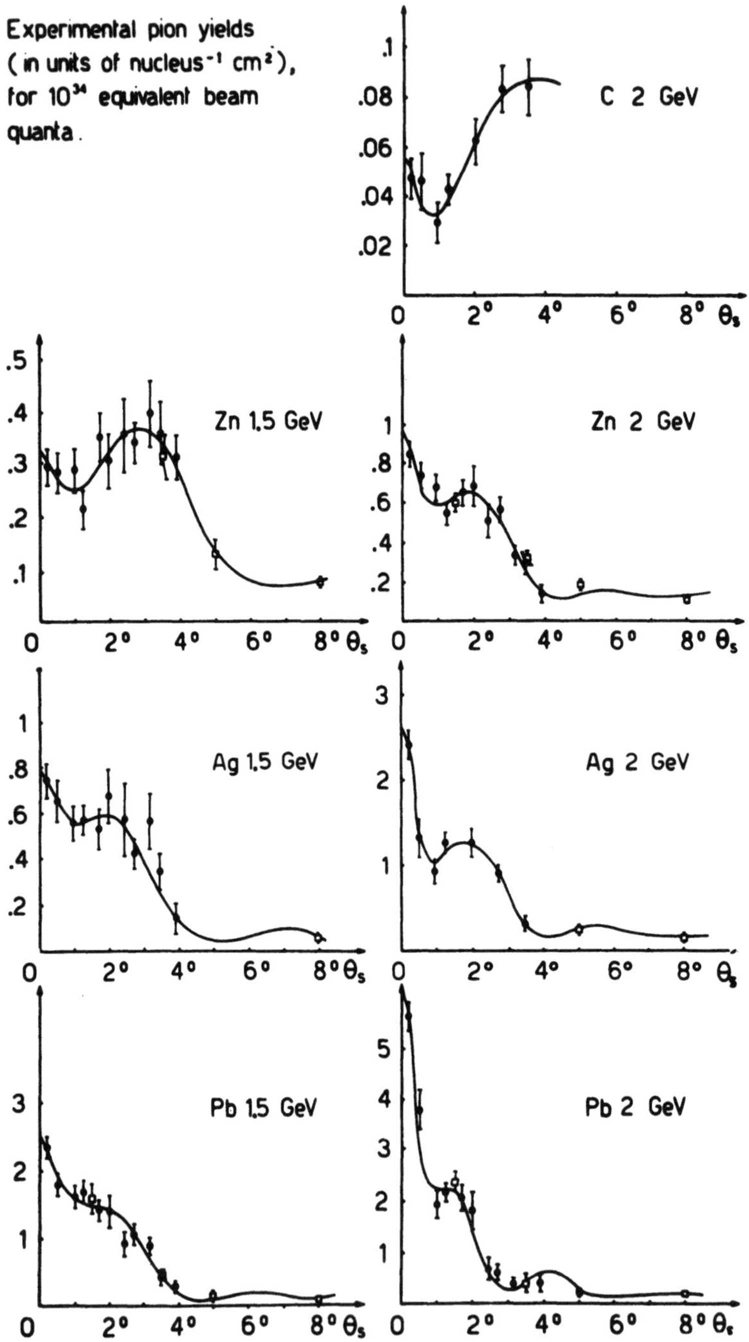

Fig. 16. Angular distributions for π^0 photoproduction

cleon. This reduces the overall cross-section for nuclear photoproduction (for instance this term was found to be negligible in the measurement of the η lifetime) and allows a clean prediction of its angular behaviour.

The last measurement of the π^0 lifetime has been performed this year at DESY by the Bonn-Pisa group [4], photoproducing π^0's on C, Zn, Ag, Pb at energies of 1,5 and 2 GeV. The photons from neutral pions decay were detected by means of the same apparatus used to measure the ω photoproduction cross-section and described in Fig. 11. The yield due to Primakoff effect was separated from the nuclear coherent production by observing the different dependence of the two cross-sections on the energy, the angle and the atomic number of the target nuclei. A summary of these dependences is given in Table 4.

The experimental angular distributions are shown in Fig. 16. The peaks at small angles show the evidence for the Primakoff effect and can be clearly distinguished from the broader bumps characteristic of the nuclear coherent production; the relative importance of the two terms is well accounted for by the full lines that represent the results of best fit analysis performed for each element separately. From each angular distribution a value of the lifetime was derived. These values, plotted in Fig. 17 as function of A, agree among themselves and do not show

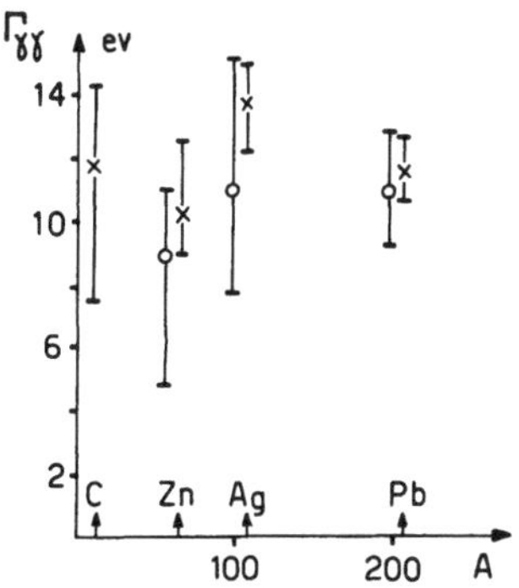

Fig. 17. Values of $\Gamma_{\pi^0 \to \gamma\gamma}$ as determined separately for each element and for each energy

any dependence on the energy and on the atomic number of the nucleus. This guarantees that the nuclear physics problems entering in the calculation of the nuclear coherent cross-section and of the interference term do not affect the determination of the π^0 lifetime. The final result, averaged on all nuclei, is

$$\tau_{\pi^0} = (0.56 \pm 0.06) \cdot 10^{-16} \text{ sec}$$

which represents the most accurate measurement performed up to the present days.

References

1. A survey of the selection rules characteristic of the coherent production can be found in: *Bemporad, C.*: Lecture delivered at the Summer Institute on Diffractive Processes, Montreal, Canada 1969.
2. *Bellettini, G., et al.*: Nuovo Cimento **90** A, 1139 (1965).
3. *Bemporad, C., et al.*: Phys. Letters **25** B, 380 (1967).
4. *Bellettini, G., et al.*: Nuovo Cimento X, **66** A, 243 (1970).
5. A review of the experimental determinations of the leptonic decay rates of vector bosons was given by *Ting, S. C. C.*: In: Proceeding of the 14th International conference on high Energy physics, Vienna 1968, p. 43.
6. *Silverman, A.*: Vector Meson Photoproduction. Proceedings of the 4th International Symposium on Electron and Photon Interactions at high energies, Liverpool 1969, p. 69.
7. *Asbury, J. G., et al.*: Phys. Rev. Letters **19**, 865 (1967), **20**, 227 (1968).
8. *McClellan, G., et al.*: Phys. Rev. Letters **22**, 377 (1969).
9. *Bulos, F., et al.*: Phys. Rev. Letters **22**, 490 (1969).
10. *Alvensleben, H., et al.*: Phys. Rev. Letters **24**, 786 (1970), **24**, 792 (1970); Nucl. Phys. B **18**, 333 (1970).
11. *McClellan, G., et al.*: Preprint submitted to the Kiev Conference 1970.
12. *Behrend, H. J., et al.*: Phys. Rev. Letters **24**, 336, 1970.
13. *Collard, H. R., Elton, L. R. B., Hofstädter, R.*: In: Landolt-Börnstein tables. New Science Group 1, Vol. 2. Berlin-Heidelberg-New York: Springer 1970.
14. *Alvensleben, H., et al.*: DESY preprint 70/40, 1970, and Phys. Rev. Letters **25**, 1377 (1970).
15. *Koelbig, K. S., Margolis, B.*: Nucl. Phys. B **6**, 85 (1968).
16. *Bauer, T.*: Ph. D. thesis. Cornell University, 1970.
17. *Bingham, H. H., et al.*: Phys. Rev. Letters **24**, 955 (1970).
18. *Behrend, H. J., et al.*: Phys. Rev. Letters 1246 (1970).
19. *Wolf, G.*: Phys. Rev. **182**, 1588 (1969).
20. *Braccini, P. L., et al.*: DESY Preprint 70/33 (1970); Nucl. Phys. B **24**, 173 (1970).
21. The models differ in the evaluation of the form factor reducing the incoherent cross section at small momentum transfers. The first two models are due to *Trefil, J. S.*: Phys. Rev. **180**, 1379 (1969) and to *Engelbrecht, C. A.*: Phys. Rev. **133** B, 988 (1964). The third model accounts for this reduction using the t-dependence of the cross section as measured in π^+ and π^- photoproduction on nuclei by *Bojarski, A. M., et al.*: Phys. Rev. Letters **23**, 1343 (1969), while the forth one simply sets this form factor equal to 1.
22. *Barbaro Gualtieri, A., et al.*: Rev. Mod. Phys. **42**, 87 (1970).
23. *Augustin, J. E., et al.*: Phys. Letters **28** B, 503 (1969).
24. For the theory of Primakoff effect see *Primakoff, H.*: Phys. Rev. **81**, 899 (1951); *Morpurgo, G.*: Nuovo Cimento **31**, 569 (1964).
25. *Ballam, et al.*: Phys. Rev. Letters **24**, 1364 (1970).

Dr. *L. Foà*
Istituto Nazionale di Fisica Nucleare
Pisa/Italy

Current Amplitudes in Dual Resonance Models

M. ADEMOLLO

Contents

1. Introduction

Soon after the dual resonance model (DRM) has proved so successful in describing certain features of particle scattering and so promising as a theory of strong interactions, a number of attempts have been made in order to extend this model to include the electromagnetic and weak interactions of hadrons. Unfortunately these attempts have reached only a partial success on the way of constructing an amplitude satisfying all the requirements of analyticity, factorization, divergence conditions and duality. Therefore a satisfactory theory of e.m. processes, in the frame of DRM, is not available until now, and the construction of such a theory, if even possible, seems to be an extremely difficult task. The purpose of this talk is to give an account of the work which has been done in this direction during the last two years.

1.1. The Dual Resonance Model

First of all, let me say a few words about DRM for strong interactions. I do not intend to enter here into the technical details, for which I refer to the numerous review papers [1] existing in the literature. I would like just to recall the basic properties of this model.

In the commonly used DRM a scattering amplitude is written down for an arbitrary number of identical scalar isoscalar mesons. Such an amplitude has an analytic expression which has the following four properties:

a) All singularities are poles corresponding to the exchange of zero-width resonances.

b) Asymptotic Regge behavior.

c) Crossing Symmetry.

d) Planar duality.

According to properties a) and b), the amplitude can be expressed, in a given channel, as a sum of poles in the corresponding energy variable. The resonances contributing to these poles lie on an infinite family of straight Regge trajectories, with a leading trajectory $\alpha(s) = a + bs$, on which also the external scalar mesons lie, and a sequence of daughter trajectories parallel to the leading one, spaced by one unit and with increasing degeneracy. The resulting spectrum of hadrons, which is needed to accomplish the full factorization of the amplitude, increases exponentially with the energy. Besides resonances, however, also ghosts appear, i.e. states with negative norm.

We observe that the hypothesis that the amplitude is expressible as a sum of poles in a given channel, implies that also the asymptotic Regge behavior, or the resonance poles exchanged in the crossed channels, be generated by the direct channel poles. This is in fact the property of duality, for which the two descriptions in terms of resonances or in terms of Regge poles are completely equivalent. This situation is quite different from any interference model, in which the two things have to be added together. However we shall use duality in a more specific way, which is called planar duality. According to this principle, the full N-point amplitude is the sum of $(N-1)!/2$ terms, each one of them corresponding to a planar graph with the external particles in a given cyclic order. All graphs differing for a cyclic or anticyclic permutation of the external lines correspond to the same term. Each term may have singularities associated with the internal lines of any possible tree graph, and all possible tree graphs (summed over the corresponding pole contributions) are equivalent. For example, in the simplest case of the 4-point function, the amplitude is given by the well known Veneziano formula [2]. The full amplitude $A(s, t, u)$ is the sum of three terms $A(s, t)$, $A(s, u)$ and $A(t, u)$. The term $A(s, t)$ is given by the Euler Beta function $B(-\alpha(s), -\alpha(t))$ and exibits simple poles in the physical regions of s and t channels, and similarly for $A(s, u)$ and $A(t, u)$. Of course planar duality also implies crossing symmetry.

The basic model described above can be easily generalized to include external particles with nonzero spin and isospin. Amplitudes for spinning mesons may be obtained by taking the residue on the corresponding poles in a tree graph with a larger number of scalar particles. We thus obtain a complete bootstrap in which hadrons of the same family may appear as external or internal particles. Furthermore, amplitudes where all external and internal particles have isospin 0 or 1, may be obtained [3] by multiplying each individual term in the dual decomposition of the full amplitude by a corresponding isospin factor.

The model we have presented is clearly not unitary and it might be thought as the analogous of the Born term in a perturbative expansion of a unitary S matrix. Higher order dual graphs, containing one or more loops, have been studied and a renormalization procedure has been devised [4] to eliminate divergences. However the evaluation of finite corrections is still an open problem.

The DRM represents an idealized world of unphysical hadrons. Amplitudes for physical particles like pions, nucleons etc. have also been studied by many authors and a number of 4-body and 5-body reactions have been fit with physical trajectories and with considerable success. However the extension of these amplitudes to the case of many particles poses several theoretical problems and in every case which has been considered some of the simple properties of the scalar meson case (like factorization or Regge behavior or even meromorphism) have been lost.

In this situation a detailed comparison with experiments seems premature and we may hope that our idealized scheme reproduces at most the gross features of the experimental facts. On the other hand the properties of the DRM are very attractive from a purely theoretical point of view and deserve to be studied by themselves.

1.2. Extension to Currents

It is natural to ask whether this model can be extended to include currents. Of course we have to limit ourselves to the case of vector currents, because axial currents would require, by PCAC, amplitudes for pions satisfying the Adler condition, and this is beyond the limits of the present model.

The construction of dual current amplitudes is also suggested by the following arguments. First, in a scheme where hadronic amplitudes are dominated by narrow resonances, it is natural to expect that also the singularities in q^2 (where q is the momentum of the current) are dominated by narrow resonances. The existence in the hadronic spectrum of an infinite number of vector mesons makes it possible in principle to construct a generalized vector meson dominance model with many free parameters.

Secondly, the electromagnetic form factors decrease rapidly for large spacelike q^2 and presumably satisfy unsubstracted dispersion relations. Therefore the current amplitudes can be written as sums of poles in q^2 as well as in the energy variables.

A third and more interesting argument comes from scaling law in deeply inelastic electron scattering. It has been recently shown by *Bloom* and *Gilman* [5] that if we consider the structure functions $W_1(\omega', t)$ and $\nu W_2(\omega', t)$ as functions of the modified variable $\omega' = -(\nu + M^2)/q^2$ $(\nu = (s-u)/2)$, the scaling law manifests itself even at relatively low energies, in the sense that the smooth curve obtained for high values of ν mediates the wavy curve obtained for values of ν in the resonance region. Although the kinematic situation is different, this property is strongly reminiscent of what we had in finite energy sum rules for strong interactions, where the Regge pole term mediates the resonances contribution, and it may suggest that current amplitudes actually obey duality.

We now come to our central problem of constructing a model for current amplitudes. Specifically, we require this model to satisfy the following conditions:

(i) All singularities of the amplitude are simple poles in q^2 and in energy variables.

(ii) Factorization, in the sense that the hadronic spectrum in any channel be consistent with DRM and that the purely hadronic amplitudes which can be factorized out of the current amplitude coincide with the corresponding amplitudes of DRM.

(iii) Good behavior for large q^2.

(iv) Divergence conditions, coming from CVC and current algebra.

(v) Planar duality.

As we said at the beginning, a model satifying all these requirements has not been given so far, and we do not even know if it is possible. The models which have been proposed can be divided into two classes:

I – Phenomenological models, which satisfy conditions (i), (iii) and (v), sometimes (iv), but have very bad factorization properties.

II – Factorizable models, which enforce conditions (ii) and (iv), but have bad behavior for large q^2.

In the next section we discuss some general properties that current amplitudes should have and the implications of the above conditions which are independent of any specific model. In the last two sections we present a brief summary of the two classes of models which have been proposed.

2. General Properties of Dual Current Amplitudes

We want to discuss in some detail some of the general properties that current amplitudes are expected to possess, and which follow from factorization, duality and CVC or are suggested from experiment or field theory models. For this discussion we follow the beautiful analysis of *Brower* and *Weis* [6].

For simplicity we shall only consider in the following amplitudes for N spinless hadrons and one or two vector currents. We call $V^{\mu}(q)$ the single-current amplitude and $M^{\mu\nu}(q_1, q_2)$ the two-current amplitude.

2.1. Factorization

In the narrow resonance approximation our amplitudes will have pole singularities in any channel. We also require convergent asymptotic behavior in each variable, so that the amplitude can be expressed as a sum of poles. Furthermore, the spectrum of hadrons which appear in any channel must be the same as in DRM. Of course in the channel of a current only vector mesons can contribute. The different situations which may arise in factorization are illustrated in Fig. 1 for the case of one current and in Fig. 2 for the case of two currents. In this last case

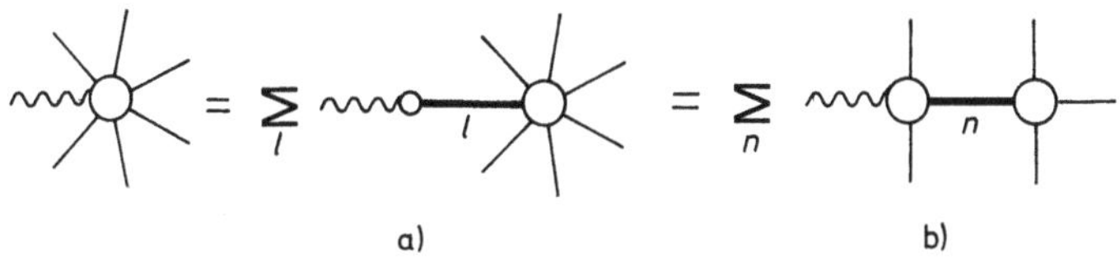

Fig. 1

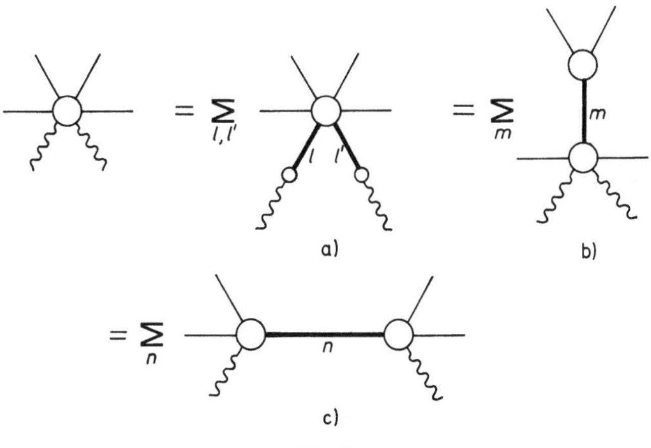

Fig. 2

there are two different possibilities of factorization in a multiparticle channel: the one of Fig. 2b, with the two currents on the same side, which is called linear factorization, and the one of Fig. 2c, with the currents on the two different sides, which is called quadratic factorization.

I want to emphasize here the crucial importance of quadratic factorization. In fact we see that each factorized amplitude of Fig. 2c is the same as the one-current amplitude of Fig. 1b, and this means that the two-current amplitude is entirely determined by the one-current amplitude. This fact of course poses very strong restrictions on the one-current amplitude, in order that the two-current amplitude satisfies e.g. the divergence conditions.

2.2. Large q^2 behavior

Now we want briefly discuss the behavior of one and two-current amplitudes for large q^2.

Concerning form factors, there are many theoretical reasons to believe that in a model like the DRM where particles are infinitely composite, they should fall exponentially. The arguments which can be brought in favour of this belief are the following:

a) Mandelstam conjecture [7], which is based on the extrapolation of the following result: in a two-body elastic scattering, if the phase shift asymptotically approaches $n\pi$, then the Omnès equation for the form factor gives a solution which goes like $(q^2)^{-n}$.

b) A more precise argument comes from current algebra plus Regge behavior for Compton scattering. Consider the local current algebra relation

$$q_{1\mu} M^{\mu\nu}(q_1, q_2) = V^\nu(q_1 + q_2), \tag{1}$$

and take $|t|$ very large ($t = (q_1 + q_2)^2$). If there are no fixed poles in the s channel (as we also assume), the left-hand side behaves like a superposition of powers $|t|^{\alpha(s)-n}$. But since the right-hand side does not depend on s, the power on the left must cancel and what remains must fall faster than any power.

c) Harte's bootstrap model [8] for infinitely composite particles, which gives a form factor $F(q^2) \to \exp[-(-q^2)^{1/4}]$.

d) Field theory models [9], which give for a particle considered as a bound state of n elementary particles a form factor going like $(q^2)^{-n}$.

Therefore, extending the above arguments, we expect that the single-current amplitudes should fall exponentially for $q^2 \to -\infty$.

For the two-current amplitudes, there are two interesting limits when both q_1^2 and q_2^2 become very large. The Bjorken limit expresses the amplitude $M_{ab}^{\mu\nu}$ as an asymptotic expansion in powers of $1/q_0$ in terms of equal time commutators of the vector current with its time derivatives. Defining $q=(q_1-q_2)/2$ and $\Delta=q_1+q_2$ and taking $|q_0|\to\infty$ with $\boldsymbol{q}, \Delta$ and the hadronic momenta fixed we have:

$$\begin{aligned} M_{ab}^{\mu\nu} \to & -\frac{i}{q_0}\int d^3x\, e^{i\boldsymbol{q}\cdot\boldsymbol{x}}\langle\alpha|\left[V_a^\mu\left(\frac{\boldsymbol{x}}{2},0\right), V_b^\nu\left(-\frac{\boldsymbol{x}}{2},0\right)\right]|\beta\rangle \\ & -\frac{1}{q_0^2}\int d^3x\, e^{i\boldsymbol{q}\cdot\boldsymbol{x}}\langle\alpha|\left[\dot{V}_a^\mu\left(\frac{\boldsymbol{x}}{2},0\right), V_b^\nu\left(-\frac{x}{2},0\right)\right]|\beta\rangle+\cdots \end{aligned} \tag{2}$$

where a and b are isospin indices and α and β represent hadronic states. The terms of this expansion correspond to fixed poles in the two-current channel, and we expect that these terms will converge for the fixed poles dominating on the Regge behavior.

Another interesting limit is given by $q_1^2=q_2^2=q^2\to-\infty$ with $\omega=-\nu/q^2$ and t fixed ($\nu=(s-u)/2$ and $t=(q_1+q_2)^2$) for Compton scattering, which is related to deeply inelastic electron scattering. Both theory [10] and experiment [11] suggest that typical invariant amplitudes behave like

$$M_i\to(-q^2)^{-n_i}f_i(\omega,t)\,, \tag{3}$$

where n_i is a small integer. In Regge behaved models we also expect

$$\mathrm{Im}\, f_i(\omega,t)\xrightarrow[\omega\to\infty]{}\omega^{\alpha(t)-r_i}. \tag{4}$$

We must say that experiment actually suggests this kind of behavior only for the diffractive contribution, and it could well be that the contribution of the dual trajectories vanishes more rapidly than in (3).

2.3. Duality and Divergence Conditions

Here we want to analyze the consequences of the general conditions (iv) and (v) of Section 1.2, namely planar duality and divergence conditions.

We first consider the case of one current. According to planar duality the full amplitude $V^\mu(q)$ is the sum of terms corresponding to all possible cyclic orderings of the external lines

$$V^\mu(q)=\sum_{i,P}V_{i,P}^\mu(q)\,, \tag{5}$$

where P specifies the permutation of the hadrons $(1,2,\ldots,N)$ and i means that the current is between the hadrons $P(i-1)$ and $P(i)$. For the moment we neglect isospin.

Now the conservation of the vector current implies that

$$q_\mu V_{i,P}^\mu(q)=0\,, \tag{6}$$

i.e. each term in the dual decomposition (5) must be individually divergenceless. In fact different terms in (5) have singularities in different channels which could not compensate otherwise. Eq. (6) is a strong consequence of duality, and we may remark that this situation is quite different from the case of Feynman graphs, where gauge invariance requires that only the sum of all possible graphs of the same order must be divergenceless.

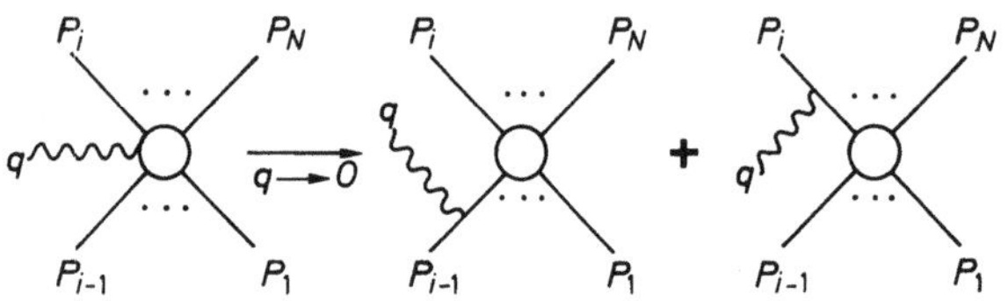

Fig. 3

It is interesting to consider the limit $q \to 0$. In this limit the dominant contribution to $V^{\mu}_{i,P}$ comes from the external line insertions (ELI) (see Fig. 3) and is given by (for simplicity we consider the fundamental permutation of hadrons and we drop the index P)

$$V_i^{\mu}(q) \xrightarrow[q \to 0]{} Q_i^- \frac{2p_{i-1}^{\mu} + q^{\mu}}{(q+p_{i-1})^2 - m^2} B_N + Q_i^+ \frac{2p_i^{\mu} + q^{\mu}}{(q+p_i)^2 - m^2} B_N, \tag{7}$$

where Q_i^- and Q_i^+ are the charges associated with the lines p_{i-1} and p_i, m is the scalar meson mass and B_N is the hadronic N-point amplitude. Condition (6) gives now

$$Q_i^- = -Q_i^+ = Q_i, \tag{8}$$

so that V_i^{μ} takes for $q \to 0$ the form

$$V_i^{\mu}(q) \xrightarrow[q \to 0]{} Q_i \left(\frac{2p_{i-1}^{\mu} + q^{\mu}}{(q+p_{i-1})^2 - m^2} - \frac{2p_i^{\mu} + q^{\mu}}{(q+p_i)^2 - m^2} \right) B_N. \tag{9}$$

We also observe that the total charge of the hadron i is $e_i = Q_{i+1} - Q_i$, from which the relation of charge conservation $\sum_i e_i = 0$ follows automatically.

We are thus led quite naturally to the duality diagrams of *Harari* [12] and *Rosner* [13] (see Fig. 4) where in the amplitude V_i^{μ} the current interacts with a single quark line through the charge Q_i. This picture is better understood when we introduce isotopic spin. If external mesons are restricted to $I = 0,1$, they can be labelled by a quark index α and an antiquark index β. Thus the isovector current amplitude, with isospin component a, can be labelled as $V^{\mu,\beta_1 \ldots \beta_N}_{a,\alpha_1 \ldots \alpha_N}$ and similarly for each term of the dual decomposition (5). Each tensor amplitude can then be decomposed into a sum of isospin invariant amplitudes, each one multiplied by a covariant of the form $\delta_{\alpha_1}^{\beta_{k_1}} \ldots (\tau_a)_{\alpha_j}^{\beta_{k_j}} \ldots \delta_{\alpha_N}^{\beta_{k_N}}$. Each δ can be graphically represented by a quark line, e.g. $\delta_{\alpha_i}^{\beta_j}$ is represented by a quark line entering with meson i and coming out with meson j. Each isospin invariant amplitude has the dual decomposition (5) and each term satisfies Eq. (6). For $q \to 0$, in the amplitude V_i^μ the current couples through τ_a to a single quark line passing between p_{i-1} and p_i.

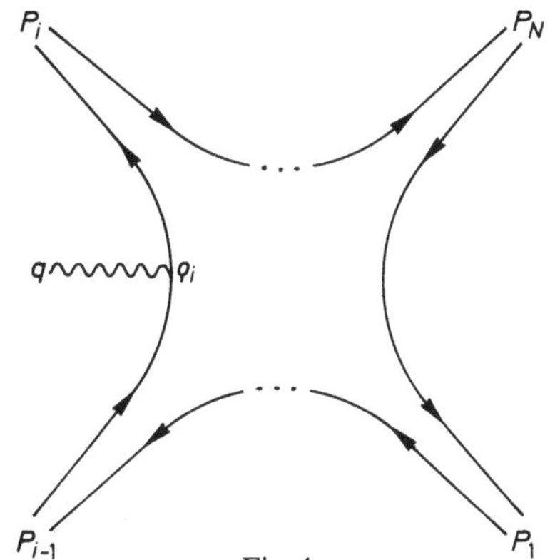

Fig. 4

In general, however, in such isospin decomposition channels with $I > 1$ may occur. If we want to avoid these exotic channels, we must say that in the dual decomposition of the full isotensor amplitude each term has only one isospin invariant non-zero amplitude, i.e. the one corresponding precisely to the quark structure of Fig. 4. By using cartesian components for the hadron isospin we have the dual decomposition

$$V_a^\mu(q) = \sum_{i,P} \mathrm{Tr}\left[\tau_{P(1)} \cdots \tau_{P(i-1)} \frac{\tau_a}{2} \tau_{P(i)} \cdots \tau_{P(N)}\right] V_{i,P}^\mu(q)\,, \tag{10}$$

where $\tau_k (k = 1, \ldots, N)$ stands for $\tau_{\mu_k} (\mu_k = 0, 1, 2, 3)$, $\mu_k = 0$ for $I_k = 0$ and $\tau_0 = 1$. This result is quite analogous to the one by *Chan* and *Paton* [3] for strong interactions. In (10) $V_{i,P}^\mu(q)$ is a planar isospin invariant amplitude, and its limit for $q \to 0$ is analogous to Eq. (9) with $Q_i = 1$.

In conclusion we see that planar duality, in connection with CVC and isospin invariance, requires a quark-like structure for the internal symmetry variables. We also observe that the existence of ELI contributions of the form (9), together with vector meson dominance, strongly suggests that the current coupling is mediated by (one or more) universally coupled vector mesons.

We now consider the case of two currents. First of all we have to state which are the divergence conditions for the full amplitude. These are better expressed for the isospin symmetric and antisymmetric parts, respectively $M^{\mu\nu}_{(a,b)}(q_1, q_2)$ and $M^{\mu\nu}_{[a,b]}(q_1, q_2)$ where a and b are isospin indices. In the general case the divergence conditions can be given in the limit $q_1 \to 0$ and are the following

$$q_{1\mu} M^{\mu\nu}_{[a,b]}(q_1, q_2) \to i\varepsilon_{abc} V^{\nu}_c(q_2), \tag{11a}$$

$$q_{1\mu} M^{\mu\nu}_{(a,b)}(q_1, q_2) \to 0. \tag{11b}$$

Eq. (11a) follows essentially from the charge-current commutation relation $[Q, V] = V$, while Eq. (11b) follows from gauge invariance for electromagnetic currents [14]. If the local current algebra relation

$$\left[V^0_a\left(\frac{x}{2}\right), V^{\nu}_b\left(-\frac{x}{2}\right)\right] \delta(x_0) = i\varepsilon_{abc} V^{\nu}_c(0)\, \delta(x) \tag{12}$$

exists, then much stronger divergence conditions can be given, for arbitrary momenta q_1 and q_2:

$$q_{1\mu} M^{\mu\nu}_{[a,b]}(q_1, q_2) = i\varepsilon_{abc} V^{\nu}_c(q_1 + q_2), \tag{13a}$$

$$q_{1\mu} M^{\mu\nu}_{(a,b)}(q_1, q_2) = 0. \tag{13b}$$

Relations analogous to (11) and (13) hold of course by exchanging the role of q_1 and q_2.

Calling now duality into play we can write the following decomposition of the full amplitude

$$\begin{aligned} M^{\mu\nu}_{ab}(q_1, q_2) = &\sum_{i \neq j} C_{ij} M^{\mu\nu}_{ij}(q_1, q_2) + \sum_i C_i M^{\mu\nu}_{ii}(q_1, q_2) \\ &+ \sum_i C'_i M'^{\mu\nu}_{ii}(q_1, q_2), \end{aligned} \tag{14}$$

where $M^{\mu\nu}_{ij}$, with $i \neq j$, corresponds to the planar amplitude where q_1 comes before p_i and q_2 before p_j in the cyclic order of the external lines; $M^{\mu\nu}_{ii}$ is the amplitude where the two currents are adjacent, q_1 precedes q_2 and this precedes p_i; $M'^{\mu\nu}_{ii}$ is the analogous where q_2 precedes q_1. The coefficients C_{ij}, C_i and C'_i, in the case where exotic channels are absent, are traces of products of τ matrices in the same order of the external lines, similarly to Eq. (10). In Eq. (14) for simplicity we have taken the hadron momenta in the order $p_1, p_2, \ldots, p_N$ and a sum over the hadron permutations should be understood. We observe that by Bose statistics $M'^{\mu\nu}_{ii}(q_1, q_2) = M^{\nu\mu}_{ii}(q_2, q_1)$.

We now consider the divergence conditions for the individual planar amplitudes. The terms $M^{\mu\nu}_{ij}$, with $i \neq$ j, have two ELI each in the limit $q_1 \to 0$, whose divergences may compensate like in the case of a single current. Therefore these terms may be taken individually divergenceless. The terms with adjacent currents, on the contrary, have only one ELI each and therefore have non-zero divergence, which is related to Eqs. (11) or (13). In conclusion we have

$$q_{1\mu} M_{ij}^{\mu\nu}(q_1, q_2) = 0, \quad (i \neq j) \tag{15a}$$

$$q_{1\mu} M_{ii}^{\mu\nu}(q_1, q_2) = V_i^\nu(q_1 + q_2), \tag{15b}$$

$$q_{1\mu} M_{ii}'^{\mu\nu}(q_1, q_2) = -V_i^\nu(q_1 + q_2). \tag{15c}$$

These relations hold for arbitrary q_1 and q_2 in the case of local current algebra and in the limit $q_1 \to 0$ in the general case.

2.4. Fixed Poles

There is no need of fixed power behavior in single-current amplitudes provided that form factors are vanishing for $q^2 \to -\infty$. In fact it has been proved by *Dashen* and *Frautschi* [15] that pure Regge behavior for current amplitudes is a consequence of Regge behavior for hadronic amplitudes, unitarity and unsubstracted dispersion relations in q^2. On the other hand this is consistent with the result of field theory models [16], in the limit of infinitely composite hadrons.

On the contrary, as is well known, fixed powers must be present in the two-current amplitudes, due to the divergence conditions. This can be immediately seen as follows. Consider e.g. the planar amplitude $M_{ii}^{\mu\nu}(q_1, q_2)$, call t the channel of the two currents and s any channel dual to this one. We can use quadratic factorization of Fig. 2c to express the amplitude as a sum of poles in s. Now the divergence $q_{1\mu} M_{ii}^{\mu\nu}(q_1, q_2)$ cannot have any such pole because of CVC and so it will be a polynomial in s. Actually it is a non-zero constant, and precisely $V_i^\nu(q_1 + q_2)$ according to Eq. (15b). Therefore, without entering into details, the amplitude itself must have some term which behaves like s^{-1}. This kind of asymptotic behavior can be related to a fixed pole or a Kronecker δ singularity in the angular momentum plane of the t channel.

We want to remark an interesting consequence of duality in this connection. In fact, by Eqs. (15), only the planar amplitudes with adjacent currents must have the fixed power behavior, in any variable dual to t. Therefore only the t channel must have the fixed pole. Once again, this is consistent with field theory models.

Finally, we may observe that the presence of fixed poles implies that certain sums on intermediate states are not uniformly convergent. Consider for example the amplitude $M_{ii}^{\mu\nu}$ expressed as a sum over the vector meson poles as in Fig. 2a. Each term of the sum has no fixed power for $s \to \infty$, while the entire sum does have it. Therefore the limit $s \to \infty$ cannot be exchanged with the sum. Another example is given by the current algebra sum rule for the form factor

$$\frac{1}{\pi} \int \mathrm{Im}\, A(s, t, q_1^2, q_2^2)\, \mathrm{d}s = \sum_n R_n(t, q_1^2, q_2^2) = F(t)\,. \tag{16}$$

Each term of the sum is expected to be rapidly vanishing for q_1^2, $q_2^2 \to -\infty$, while the entire sum is independent of q_1^2 and q_2^2.

These rather peculiar features seem very difficult to obtain in a dual factorizable model.

3. Phenomenological Models

The early models which have been proposed [17–22] for dual current amplitudes were essentially concerned with analytic properties in general, like the correct pole structure in all channels, including currents, and asymptotic behavior. Divergence conditions were enforced only in particular cases, while factorization was never satisfied, except for the leading trajectory. In absence of factorization, the construction of current amplitudes is obviously nonunique and in fact the functions which have been proposed are rather different. However they are all obtained by suitable modifications of the hadronic N-point function and possess several features in common. We believe that these amplitudes are especially interesting from the phenomenological point of view, but it may well be that some useful hint can also be obtained for the solution of the full problem.

To be a little more specific, we shall consider the model of Ref. [20]. We remember that the hadronic N-point function described in Section 1.1 is given by

$$B(p_1, \ldots, p_N) = \int_0^1 \mathrm{d}u_1 \ldots \mathrm{d}u_{N-3}\, J^{-1}(u_i) \prod_{i,j} u_{ij}^{-\alpha_{ij}-1}\,. \tag{17}$$

where $u_1, \ldots, u_{N-3}$ are the integration variables, which are associated to the internal lines of a multiperipheral tree graph; $J^{-1}(u_i)$ is a suitable function which makes the integration volume invariant under cyclic permutations; u_{ij} are variables, functions of the u_i, associated to all possible channels, and $\alpha_{ij} = a + b(p_i + \cdots + p_j)^2$ are the corresponding trajectory functions. A fundamental property of (17) is its multi-Regge behavior, i.e. for asymptotic values of any energy variable we have

$$B \xrightarrow[\alpha_{ij} \to -\infty]{} \sum_{\bar{\alpha}} (-\alpha_{ij})^{\bar{\alpha}}\,, \tag{18}$$

where $\bar{\alpha}$ is the trajectory of any channel which is dual to (i, j).

The function (17) is modified as follows in order to describe current amplitudes. For each current we introduce two lepton lines and we consider the amplitude for the N hadrons plus the leptons (see Fig. 5). This amplitude is then written as a product of the hadronic current amplitude times the lepton currents. The hadronic current amplitudes have a structure similar to (17), except that for channels corresponding to one lepton plus r hadrons we introduce a fictitious constant trajectory $\alpha_{ij} = -\gamma$, while for hadronic channels, including currents, we use the universal trajectory function $\alpha(s) = a + bs$. In the case of two currents there is also one channel containing one lepton associated to one current and one lepton associated to the other current, and for this channel we also take a constant trajectory $-\gamma_1$. The model of Ref. [21] is essentially coincident with this one, except that it assumes $\gamma = \gamma_1 = 1$ and it only refers to some non specified invariant amplitude for each process.

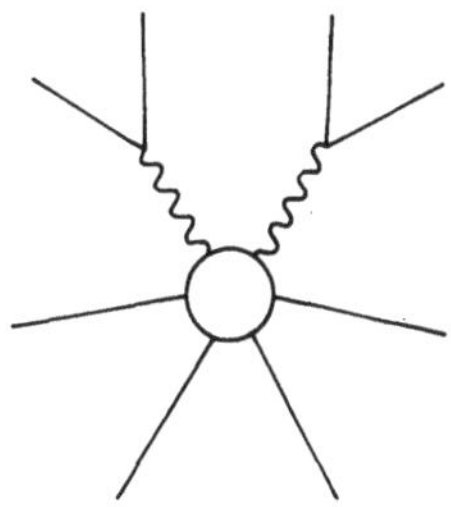

Fig. 5

Due to the asymptotic behavior (18) of the dual amplitudes, the presence of constant trajectories reflects in the appearance of fixed power behaviors, in addition to Regge, in every channel.

In general, the main properties of these amplitudes are the following:

a) correct pole structure in the energy variables and in q^2;

b) fixed poles of the multiplicative type at nonsense points in every channel, corresponding to fixed power behavior in energy variables and q^2;

c) factorization on the first vector meson pole in the currents, in the sense that the residue at the pole $\alpha(q^2) = 1$ coincides with the strong amplitudes of DRM;

d) factorization on the leading trajectory in every channel;

e) divergence conditions are not satisfied in general.

Let us see some particular examples. The scalar meson form factor takes the simple form

$$F(t) \propto B[1 - \alpha(t), \gamma - 1], \tag{19}$$

and for $|t| \to \infty$ behaves like $|t|^{-\gamma+1}$.

The electroproduction amplitude of Fig. 6 is given by

$$V_\mu = \int_0^1\int_0^1 \frac{dx_1 dx_2}{1-x_1x_2} x_1^{-\alpha(q^2)} x_2^{-\alpha(s)-1} \left(\frac{1-x_2}{1-x_1x_2}\right)^{-\alpha(t)-1} (1-x_1)^{\gamma-1}$$
$$\cdot \left[(p_1+p_2+p_3)_\mu \frac{1-x_2}{1-x_1} + (p_1-p_2-p_3)_\mu \frac{x_2}{1-x_1x_2}\right]. \tag{20}$$

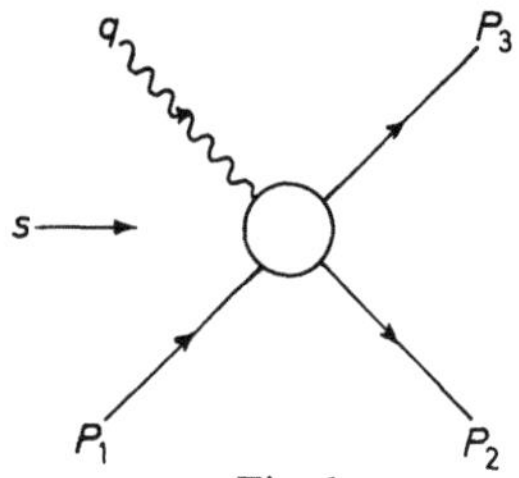

Fig. 6

We can see that the low lying poles are at the right positions for the different helicity amplitudes and the same is true for the asymptotic behaviors. Here, however, besides the good Regge poles there are also fixed poles, as seen in general before. Furthermore it turns out that the amplitude (20) satisfies the divergence condition $q^\mu V_\mu = 0$. This result, however, is not true for multiple production processes.

Analysis of Compton scattering shows that current algebra is not satisfied. By taking for the fictitious trajectories $\gamma_1 = 1, \gamma > 1$ the relevant amplitude has the required behavior s^{-1}, but the residue of the fixed pole does not coincide with the form factor (19) as it should be by current algebra, although it is a different form factor, independent of q_1^2 and q_2^2. Analogous situation is obtained for the Bjorken limit, while in the limit of deeply inelastic electron scattering the amplitude shows the scaling property.

This model has been recently improved by *Weis* [23] obtaining exponential form factors and eliminating the unwanted fixed poles. His prescription is to make for the fictitious channels the following replacement in the integrand of Eq. (17)

$$u_{ij}^{\gamma-1} \to u_{ij}^{\gamma-1} \exp(-u_{ij}^{-\delta}), \quad 0 < \delta < 1 \tag{21}$$

It is then possible to eliminate all the power behaviors in form factors and energy variables, except for the fixed pole in the channel of the two adjacent currents, which is needed by the divergence conditions.

In conclusion, many of the general properties a current amplitude should have, which have been discussed in Section 2, can be reproduced in these phenomenological models. The properties which remain not satisfied are the divergence conditions and factorization.

Concerning the inclusion of divergence conditions and current algebra, I must say that this has been attempted by several authors [17, 22, 24]. However they were concerned only with particular amplitudes (Compton scattering) and their method does not seem to have general validity.

With respect to factorization, on the other hand, amplitudes like those discussed in this Section have very bad properties. In fact it has been shown by *Freedman* [25] that the spectrum of hadrons in a given channel not only is much richer than in the strong amplitudes, but also depends on the channel itself. Therefore we can say that factorization breaks down, and in this situation it is not even possible to interpret the pole singularities as due to the exchange of resonances of definite quantum numbers.

4. Factorizable Models

A different approach to the construction of dual current amplitudes consists in the explicit use of factorization. This in fact appears as the most natural way to overcome the diseases and the ambiguities of the phenomenological models. In fact if one could know the general current-two reggeons vertex function it would be easy to write down the amplitude for any process. This problem however has not yet been solved and we are not sure that such a vertex function compatible with all requirements of Section 1.2 actually exists. This difficulty is strongly related to the difficulties which are encountered in the off-shell extrapolation of the hadronic amplitude and will probably require a deeper understanding of duality.

A first attempt to combine phenomenological amplitudes of the form discussed in the previous Section with factorization was made by *Freedman* [25]. He constructed an amplitude for one scalar current, one conserved vector current and N scalar hadrons which satisfies the expected scalar-vector current algebra. However only the scalar current has a pole-dominated structure, while the vector current is structureless, i.e. has no singularities in q^2, and therefore has a bad asymptotic behavior in q^2. Furthermore, the factorization properties of this amplitude are not the ones required at point (ii) of Section 1.2, as there are many more states than in purely hadronic amplitudes, although the inconsistency of the case of two vector currents is avoided.

A more systematic approach to factorization has been tried by *Brower* and *Weis* [26] using the operator formalism and the Ward identities of DRM.

The amplitude for $r+s$ scalar hadrons, corresponding in the s channel to the process $p_1+p_2+\cdots+p_r \to p'_1+p'_2+\cdots+p'_s$, can be written in a factorized form as $\langle p'|D(s)|p\rangle$, where $|p\rangle$ and $\langle p'|$ represent the initial and final multiparticle states and $D(s)$ is the propagator operator containing all the pole structure of the s channel.

The Ward identities of DRM originate from the existence of an operator W which annihilates any $|p\rangle$ state, i.e.

$$W(\pi)|p\rangle = 0\,, \qquad \left(\pi = \sum_i p_i\right), \tag{22}$$

$$W(\pi) = -\pi^2 + \sqrt{2}\pi\cdot a^{(1)} - \sum_{n=1}^{\infty} n a^{(n)\dagger}\cdot a^{(n)} + \sum_{n=1}^{\infty}\sqrt{n(n+1)}\,a^{(n)\dagger}\cdot a^{(n+1)}, \tag{23}$$

where $a_\mu^{(n)\dagger}$ are the creation operators which create all the states $|\lambda\rangle$ of the hadronic spectrum and $a_\mu^{(n)}$ are the corresponding destruction operators. They satisfy the commutation relation

$$[a_\mu^{(m)}, a_\nu^{(n)\dagger}] = -\delta_{mn} g_{\mu\nu}\,. \tag{24}$$

The operator W acting on any state $\langle\lambda|$ creates a so called spurious state $\langle\lambda_s| = \langle\lambda|W$, i.e. a state which does not couple to the $|p\rangle$ states. All the Ward identities of DRM are relations of the form $\langle\lambda|W|p\rangle = 0$, for all $\langle\lambda|$. For example if $\langle\lambda|$ is the ground state $\langle 0|$, we have from (23)

$$\langle 0|W(\pi)|p\rangle = \langle 0|\sqrt{2}a^{(1)}\cdot\pi - \pi^2|p\rangle = 0\,. \tag{25}$$

Now for the vector current amplitude the most natural choice is to assume

$$V^\mu(q) = F(q^2) < 0|\sqrt{2}a^{(1)\mu} + q^\mu|p\rangle\,. \tag{26}$$

In fact, in virtue of Eq. (25) ($q = -\pi$), this is automatically divergenceless. The form factor $F(q^2)$ contains all the pole structure of the current and remains here completely arbitrary.

Problems arise when we consider two-current amplitudes. In fact these should be determined by the single-current amplitude (26) and quadratic factorization. In this factorization, however, spurious state singularities will appear, which remain also in the divergence $q_{1\mu}M^{\mu\nu}$, unless we use some projection operator which eliminates spurious states. This is in fact possible, but the amplitude which is obtained does not satisfy the divergence conditions (15). Related to this fact, this amplitude also has unphysical pole singularities in the t channel, whose position depends on the current masses q_1^2 and q_2^2. Therefore we conclude that full factorization is, in this model, impossible.

Brower and *Weis* [26], however, succeeded in constructing a partially factorizable model which has the following properties:

a) factorizes on the first M trajectories;

b) has form factors falling like $(q^2)^{-M}$;

c) obeys the divergence conditions (15) of current algebra.

Without going into the details, their procedure works in the following way. Generalizing the method of Ref. [19] they write the amplitude in the form

$$M_{ii}^{\mu\nu} = M_{\mathrm{H}}^{\mu\nu} + M_{\mathrm{C}}^{\mu\nu} + M_{\mathrm{FP}}^{\mu\nu}, \tag{27}$$

where the "hadronic" amplitude $M_{\mathrm{H}}^{\mu\nu}$ contains two terms, one constructed from (26) and quadratic factorization without spurious states on the first M trajectories and one contributing only to lower trajectories. The "correction" term $M_{\mathrm{C}}^{\mu\nu}$ cancels the divergence of M_{H}, i.e. $q_{1\mu}(M_{\mathrm{H}}^{\mu\nu} + M_{\mathrm{C}}^{\mu\nu}) = 0$. The "fixed pole" term M_{FP} satisfies the exact current algebra relation

$$q_{1\mu} M_{\mathrm{FP}}^{\mu\nu}(q_1, q_2) = V^{\nu}(q_1 + q_2). \tag{28}$$

Furthermore the sum $M_{\mathrm{C}}^{\mu\nu} + M_{\mathrm{FP}}^{\mu\nu}$ has poles in s corresponding to trajectories displaced by at least M units below the leading one, so that factorization on the first M trajectories is insured by only the first term in M_{H}. However this is the only piece containing the form factors $F(q_1^2)$ and $F(q_2^2)$, and the other terms have no structure in q_i^2, so that the amplitude (27) has a bad (constant) behavior for large q_1^2 and q_2^2. Furthermore, all dual amplitudes, and especially $M_{ij}(i \neq j)$, have very bad behavior for deeply inelastic scattering, due to the multiplicative nature of the form factors. Therefore not even this model is completely satisfactory.

Another attempt to construct a fully factorizable model has been made by *Bouchiat*, *Gervais* and *Sourlas* [27]. Their model, however, essentially coincides [28] with the one of *Brower* and *Weis* discussed before and suffers from the same deseases. The merit of these authors, however, consists in having put into evidence the connection with infinite component field theory.

Recently, *Leutwyler* [29] proposed an infinite component wave equation having the same mass spectrum of DRM and which, upon quantization, gives for the Born term exactly the same amplitude of DRM. Therefore we may consider the Leutwyler model as the field theory underlying DRM. We may ask at this point which is the minimal conserved current of this model, and it turns out [27, 28] that this is precisely the one corresponding to the amplitude (26).

In conclusion, on one hand we may be happy for having a field theory which gives us a better understanding of dual amplitudes and a definite approach to currents; on the other hand we are left rather sceptic on the real possibility of constructing current amplitudes satisfying the conditions of Section 1.2. Of course this possibility exists through the introduction of non-minimal terms, corresponding also to higher oscillation modes in the amplitude (26). However a detailed investigation in this direction has not yet been made.

References

1. *Chan, H.M.:* Proc. Roy. Soc. **318**, 379 (1970).
   Jacob, M.: Proc. of the Lund Int. Conf. on Elementary Particles, Lund, 1969.
   Lovelace, C.: Proc. Conf. on $\pi\pi$ and $K\pi$ interactions, Argonne (1969), and CERN preprints TH. 1041 and TH. 1123 (1969).
   Veneziano, G.: Lecture notes for the International School of Subnuclear Physics, Erice 1970.
   Alessandrini, V., Amati, D., Le Bellac, M., Olive, D.: Dual multiparticle theory. CERN TH. 1160 (1970).
2. *Veneziano, G.:* Nuovo Cimento **57** A, 190 (1968).
3. *Chan, H. M., Paton, J.:* Nucl. Phys. B **10**, 519 (1969).
4. *Gross, D. J., Neveu, A., Scherk, J., Schwarz, J. H.:* Phys. Letters **31** B, 592 (1970); *Burnett, T. H., Gross, D. J., Neveu, A., Scherk, J., Schwarz, J. H.:* Phys. Letters **32** B, 115 (1970); *Gross, D. J., Neveu, A., Scherk, J., Schwarz, J. H.:* Phys. Rev. D **2**, 697 (1970).
5. *Bloom, E. D., Gilman, F. J.:* Phys. Rev. Letters **25**, 1140 (1970).
6. *Brower, R. C., Weis, J. H.:* Phys. Rev. **188**, 2486 (1969); see also Ref. [23].
7. *Mandelstam, S.:* Proc. of the 1966 Tokio Summer Lectures in Theoretical Physics. New York: W. A. Benjamin 1967.
8. *Harte, J.:* Phys. Rev. **171**, 1825 (1968); **184**, 1936 (1969).
9. *Ciafaloni, M., Menotti, P.:* Phys. Rev. **173**, 1575 (1968); – *Ciafaloni, M.:* Phys. Rev. **176**, 1898 (1968); – *Amati, D., Jengo, R., Rubinstein, H. R., Veneziano, G., Virasoro, M. A.:* Phys. Letters **27** B, 38 (1968).
10. *Bjorken, J. D.:* Phys. Rev. **179**, 1547 (1969); – *Abarbanel, H. D. I., Goldberger, M. L., Treiman, S. B.:* Phys. Rev. Letters **22**, 500 (1969); – *Drell, S. D., Levy, D. J., Yan, T.-M.:* Phys. Rev. Letters **22**, 744 (1969). – *Altarelli, G., Rubinstein, H. R.:* Phys. Rev. **187**, 2111 (1969).
11. For a review see e.g.: *Panofski, W. K. H.:* Proc. of the Int. Conf. on High Energy Physics, Vienna (1968); – *Lohrmann, E.:* Proc. of the Lund Int. Conf. on Elementary Particles (1969).
12. *Harari, H.:* Phys. Rev. Letters **22**, 562 (1969).
13. *Rosner, J.:* Phys. Rev. Letters **22**, 689 (1969).
14. Purely electromagnetic amplitudes contribute only to the symmetric part, say $M^{\mu\nu}_{(em)}$. For a real photon of momentum q_1 and polarization ε the amplitude is $\varepsilon_\mu M^{\mu\nu}_{(em)}$ and it must obey gauge invariance $q_{1\mu} M^{\mu\nu}_{(em)} = 0$, for $q_1^2 = 0$.
15. *Dashen, R. F., Frautschi, S. C.:* Phys. Rev. **143**, 1171 (1966).
16. *Rubinstein, H. R., Veneziano, G., Virasoro, M. A.:* Phys. Rev. **167**, 1441 (1968); – *Dashen, R., Lee, S. Y.:* Phys. Rev. Letters **22**, 366 (1969).
17. *Bander, M.:* Nucl. Phys. B **13**, 587 (1969).
18. *Sugawara, H.:* Tokio University of Education report, 1969 (unpublished).
19. *Brower, R. C., Halpern, M. B.:* Phys. Rev. **182**, 1779 (1969).
20. *Ademollo, M., Del Giudice, E.:* Nuovo Cimento **63** A, 639 (1969).
21. *Ohba, I.:* Progr. Theor. Phys. **42**, 432 (1969).
22. *Landshoff, P. V., Polkinghorne, J. C.:* Nucl. Phys. B **19**, 432 (1970).
23. *Weis, J. H.:* Preprint UCRL-19780 (1970).

24. *Brower, R. C., Rabl, A., Weis, J. H.*: Nuovo Cimento **65** A, 654 (1970).
25. *Freedman, D. Z.*: Phys. Rev. D **1**, 1133 (1970).
26. *Brower, R. C., Weis, J. H.*: Phys. Rev. **188**, 2495 (1969) and Phys. Rev. D **3**, 451 (1971).
27. *Bouchiat, C., Gervais, J. L., Sourlas, N.*: Lett. Nuovo Cimento **3**, 767 (1970).
28. *Hasslacher, B., Sinclair, D. K.*: Lett. Nuovo Cimento **4**, 515 (1970).
29. *Leutwyler, H.*: Phys. Letters **31** B, 214 (1970).

Dr. *M. Ademollo*
Istituto di Fisica Teorica dell'Università di Firenze
and Istituto Nazionale di Fisica Nucleare, Sottosezione di Firenze,
Firenze, Italy

Low Energy Photo and Electroproduction, Multipole Analysis by Current Algebra Commutators

C. VERZEGNASSI

Contents

I. Introduction

A very peculiar and interesting feature of the one-pion photo- and electroproduction processes on nucleons is that it is possible to describe them starting from two completely independent theoretical approaches. The first one is based on the dispersion theory, and its first effort to tackle the difficult multipole analysis is contained in the classical papers by *Chew, Goldberger, Low, Nambu* [1] and *Fubini, Nambu, Wataghin* [2]. Since then, almost fifteen years went by and the multipole analysis has noticeably improved [3], even if the main results contained in those papers are essentially still of value.

The second approach, recently developed by *Fubini* and *Furlan* [4], makes use of current algebra equal time commutation relations saturated in the Breit frame of two external nucleons. This method has been applied to the photo- and electroproduction by *Furlan, Paver* and *Verzegnassi* [5] and their latest result will be discussed here.

A few words of comment on the specific role of the "soft pion theorems" [6] is perhaps appropriate at this point. In our case these theorems supply expressions for some of the photo- and electroproduction invariant amplitudes, rigorously exact in the unphysical situation of a pion with vanishing four-momentum.

Starting from these theorems, there have been some efforts [7] to "extrapolate" the "soft pion" predictions to physical amplitudes, i.e. amplitudes describing a situation with the pion on the mass shell. The various extrapolation techniques essentially depend on the intuition and fancy of the authors, but a definite and unambiguous prescription has never been given. Moreover, it is easy to illustrate the kind of troubles one can meet in this type of approach by simply recalling that the main idea implied is that the result changes very slightly when passing e.g. from the point ($\boldsymbol{q}=q_0=0$) (q is the pion four-momentum) to the point ($\boldsymbol{q}=0$, $q_0=m_\pi$), that is, the physical threshold. Now, if one considers the case of the longitudinal S-wave, one realizes that the pion pole contribution, $\sim m_\pi^2/t-m_\pi^2$, is exactly vanishing in the point ($\boldsymbol{q}=q_0=0$) but largely dominates at threshold, at least for a small photon "mass". This shows that the indications given by the "soft pion" theorems can be sometimes completely meaningless.

From this point of view, the method developed by Fubini and Furlan is essentially different. This method supplies an expression for the physical amplitude from the beginning, where the pion is on the mass shell. This amplitude, in the "soft pion limit", automatically reproduces the known, exact results; however, one can in principle make use of the method ignoring completely the existence of any kind of "soft pion" theorem, treating the physical situation only.

II. Kinematics and Notation

In the usual one-photon exchange approximation, both the photo-and electroproduction processes (represented in Fig. 1) are expressed in terms of the quantity:

$$T_\mu^\alpha \equiv i\langle p_2, q^\alpha | V_\mu^{\text{e.m.}}(0)| p_1\rangle . \tag{II.1}$$

where p_2, p_1 are the final and initial momenta of the nucleon, q is the pion four-momentum and $\kappa=p_2+q-p_1$ is the four-momentum of the photon associated with the electromagnetic current.

We define the usual variables

$$\begin{gathered} W^2\equiv s=(p_1+\kappa)^2; \quad t=(p_2-p_1)^2; \quad \nu=P\cdot q; \\ P=\tfrac{1}{2}(p_1+p_2); \quad \Delta=(p_2-p_1). \end{gathered} \tag{II.2}$$

The isotopic decomposition is:

$$T_\mu^\alpha=\delta_{\alpha 3}\, T_\mu^{(+)}+\tfrac{1}{2}\,[\tau_\alpha, \tau_3]_-\, T_\mu^{(-)}+\tau_\alpha\, T_\mu^{(0)}. \tag{II.3}$$

As far as the spin decomposition is concerned, there is a wide number of possibilities. For our particular purposes we will find it convenient to make use of a slightly unusual definition*. With this aim, we first define a four-vector q':

$$q'_\mu=q_\mu-\frac{\nu}{P^2}P_\mu . \tag{II.4}$$

* We use the same metric and normalization defined in the first of Refs. [5, pg. 521–522].

This four vector has the following property:

$$q' \cdot P = 0. \tag{II.5}$$

Making use of q' we define now the amplitude T_μ as:

$$\begin{aligned} T_\mu = \bar{u}(\boldsymbol{p}_2)\,\gamma_5\,\{&P_\mu \overline{P}_1 + \hat{q}' P_\mu \overline{P}_2 + \gamma_\mu \overline{P}_3 + [\hat{q}'\gamma_\mu - q'_\mu]\,\overline{P}_4 \\ &+ \Delta\mu \overline{P}_5 + \hat{q}'\Delta_\mu \overline{P}_6 + q'_\mu \overline{P}_7 + \hat{q}' q'_\mu \overline{P}_8\}\, u(\boldsymbol{p}_1), \qquad \hat{q}' \equiv \gamma_\mu q'^\mu. \end{aligned} \tag{II.6}$$

The eight amplitudes P_i are obviously not independent, due to gauge invariance which imposes two constraints between them.

From an experimental point of view, the invariant amplitudes are not very meaningful. The relevant quantities are the multipoles (or, rather, some particular linear combinations of the multipoles), which are defined starting from the final pion-nucleon c.m.s. amplitudes $\mathscr{F}_i$.

These amplitudes, and the related multipole expansion, are the same used by *Adler* [3]. To fix the normalization, we give the expression for the c.m.s. unpolarized photoproduction cross section at threshold:

$$\frac{|\boldsymbol{\kappa}|}{|\boldsymbol{q}|}\,\frac{d\sigma}{d\Omega}(\text{th.}) = \frac{\alpha}{4\pi}\,\frac{M^2}{W^2}\,|E_{0+}|^2. \qquad \alpha = \frac{1}{137}. \tag{II.7}$$

By definition, we shall call "low energy region" the one that can be described by the terms $|\boldsymbol{q}|^0$, $|\boldsymbol{q}|^1$, $|\boldsymbol{q}|^2$, in a power series in the pion c.m.s. three-momentum $|\boldsymbol{q}|$. In the photoproduction case this could have a meaning for $|\boldsymbol{q}| \leqq \mu$ (the pion mass) which corresponds roughly to $E_\gamma^L \leqq 235$ MeV. Instead of the separate p-wave multipoles, $E_{1^+}, M_{1^\pm}$ we shall consider the more significant combinations

$$\begin{aligned} x &\equiv 3\,E_{1^+} + M_{1^+} - M_{1^-} \\ y &\equiv 2\,M_{1^+} + M_{1^-} \\ z &\equiv E_{1^+} - M_{1^+}. \end{aligned} \tag{II.8}$$

Moreover, we define

$$\begin{aligned} &\chi \equiv M_{2^+} - M_{2^-} - E_{2^+} - E_{2^-}; \qquad \psi = 2\,M_{2^+} + E_{2^+}. \\ &F_1 \equiv |\mathscr{F}_1(\vartheta = \pi/2)| = |E_{0^+} + 3\,M_{2^-} + E_{2^-} - 3\,M_{2^+} - \tfrac{3}{2}\,E_{2^+}|. \end{aligned} \tag{II.9}$$

The expression e.g. of the unpolarized differential cross section becomes then (ϑ denotes the angle between the pion and the photon directions):

$$\frac{d\sigma}{d\Omega} = \frac{|\boldsymbol{q}|}{|\boldsymbol{\kappa}|}\,\frac{\alpha}{4\pi}\,\frac{M^2}{W^2}\,[A + B\cos\vartheta + C\cos^2\vartheta]. \tag{II.10}$$

where

$$\begin{aligned} A &= F_1^2 + \tfrac{9}{2}\,z^2 + y^2 + 3\,yz + 3\,E_{0^+}\,\chi \\ B &= 2\,x\,E_{0^+} \\ C &= 3\,E_{0^+}\,\chi - \tfrac{9}{2}\,z^2 - y^2 - 3\,yz + x^2 + 15\,E_{0^+}\,\psi. \end{aligned} \tag{II.11}$$

The simple parametrization (II.10) is actually unsatisfactory if we want to describe charged pion processes for very small values of ϑ. However, if one is interested e.g. in the value $\vartheta = \pi/2$, this approximation should be meaningful.

In the electroproduction case, we may expect that a parametrization like (II.10) is still valid when $|\boldsymbol{q}| \leqq \mu$. The formulae in this case are only more complicated, involving also longitudinal multipoles, more precisely the combinations:

$$\begin{aligned} &u \equiv 4\,\mathscr{L}_{1^+} + \mathscr{L}_{1^-}; \qquad v \equiv \mathscr{L}_{1^-} - 2\,\mathscr{L}_{1^+}; \\ &F_5 \equiv |\kappa_0 \mathscr{F}_5(\vartheta = \pi/2)| = |\mathscr{L}_{0^+} + 2\,\mathscr{L}_{2^-} - \tfrac{9}{2}\,\mathscr{L}_{2^+}|\,. \end{aligned} \tag{II.12}$$

Since the current algebra approach is particularly convenient to derive expressions for the invariant amplitudes, Eq. (II.6), at threshold*, the point is how to extract from those informations the expression for the s, p and d waves (or at least for some of them) in the low energy region previously defined.

As far as the s-wave multipoles are concerned, we have the exact relations:

$$E_{0^+}(\text{th.}) = -\left[\sqrt{\frac{P^2}{m^2}}\,\overline{P}_3\right]_{(\text{th.})} \tag{II.13}$$

$$\frac{\mathscr{L}_{0^+}}{\kappa_0}(\text{th.}) = \frac{1}{2m+\mu}\,\frac{m}{\mu\sqrt{P^2}}\left\{\mu\nu\left[\overline{P}_2 + \frac{m}{P^2}\overline{P}_4\right] - \nu\left[\overline{P}_1 + \frac{\Delta\cdot q'}{2P^2}\overline{P}_4\right]\right\}_{(\text{th.})}. \tag{II.14}$$

For the p and d wave multipoles, one can verify the following relations:

$$\lim_{|\boldsymbol{q}|\to 0} \frac{2M_{1^+} + M_{1^-}}{|\boldsymbol{\kappa}|\,|\boldsymbol{q}|} = [f(\overline{P}_3, \overline{P}_4)]_{(\text{th.})} \tag{II.15}$$

$$\lim_{|\boldsymbol{q}|\to 0} \frac{E_{1^+} - M_{1^+}}{|\boldsymbol{\kappa}|\,|\boldsymbol{q}|} = [g(\overline{P}_3, \overline{P}_4, \overline{P}_5, \overline{P}_6, \overline{P}_7, \overline{P}_8)]_{(\text{th.})} \tag{II.16}$$

$$\lim_{|\boldsymbol{q}|\to 0} \frac{\mathscr{L}_{1^-} - 2\,\mathscr{L}_{1^+}}{\kappa_0|\boldsymbol{\kappa}|\,|\boldsymbol{q}|} = [h(\overline{P}_5, \overline{P}_6, \overline{P}_7, \overline{P}_8)]_{(\text{th.})} \tag{II.17}$$

$$\lim_{|\boldsymbol{q}|\to 0} \frac{M_{2^+} - E_{2^+} - M_{2^-} - E_{2^-}}{|\boldsymbol{q}|^2} = [l(\overline{P}_3, \overline{P}_4, \overline{P}_5, \overline{P}_6, \overline{P}_7, \overline{P}_8]_{(\text{th.})} \tag{II.18}$$

where f, g, h, l are certain complicated linear combinations, with coefficients depending only on κ^2, of the amplitudes at threshold.

---

* The latter is defined by the following kinematics:

$$s = (m+\mu)^2; \quad t = \frac{m}{m+\mu}(\kappa^2 - \mu^2); \quad m \equiv M_{\text{nucleon}}.$$

In the low energy region, the behaviour of the p, d waves is $\sim|\boldsymbol{q}|^{1,2}$ $(1+|\boldsymbol{q}|^2)$. Our approximation will be to parametrize the p and d wave in the low energy region as:

$$P_{\mathrm{w}}(|\boldsymbol{q}|) \simeq |\boldsymbol{q}|\,|\boldsymbol{\kappa}| \cdot \lim_{|\boldsymbol{q}|\to 0} \frac{P_{\text{wave}}}{|\boldsymbol{q}|\,|\boldsymbol{\kappa}|}\,. \tag{II.19}$$

$$D_{\mathrm{w.}}(|\boldsymbol{q}|) \simeq |\boldsymbol{q}|^2 \cdot \lim_{|\boldsymbol{q}|\to 0} \frac{D_{\text{wave}}}{|\boldsymbol{q}|^2}\,. \tag{II.20}$$

In this way, only the threshold quantities (II.13–II.18) have to be evaluated. Notice that it is unfortunately impossible to express e.g. the quantity x (Eq. II.8) by means of amplitudes at threshold. The multipoles which can be derived in such a way are those which describe the process in the configuration $\vartheta=\pi/2$. We shall limit ourselves for the moment to this particular case.

III. The Method

Our starting point is represented by the two equal time commutators:

$$[A_0^\alpha(\xi,0),\, V_\mu^{\text{e.m.}}(\vec{0},0)] = i\varepsilon_{\alpha 3\gamma} A_\mu^\gamma(\vec{0}0)\,\delta(\xi) - i g_\mu^\kappa \partial_\kappa \delta(\xi) \cdot S_{AV}^{\alpha\,3}(\vec{0},0)\,, \tag{III.1}$$

$$[D_\alpha(\xi,0),\, V_0^{\text{e.m.}}(\vec{0},0)] = i\varepsilon_{\alpha 3\gamma} D^\gamma(\vec{0},0)\,\delta(\xi)\,. \tag{III.2}$$

It has been shown in a paper by *Gross* and *Jackiw* [8] that the validity of the second commutator is a necessary and sufficient condition for the validity of the first one, i.e. for the absence of Schwinger terms proportional to higher gradients of delta functions. As to the isotopic structure of $S_{AV}^{\alpha 3}$, we will make no definite hypothesis. It could have in general a symmetric as well as an antisymmetric part.

We multiply now both sides of Eq. (III.1) by an arbitrary factor $e^{-i\boldsymbol{q}\cdot\boldsymbol{x}+i\boldsymbol{\kappa}\cdot\boldsymbol{y}}$, insert them between two one nucleon states of momentum $\boldsymbol{p}_1, \boldsymbol{p}_2$ and integrate over $\boldsymbol{x}, \boldsymbol{y}$. As a result we obtain a formal equality, whose explicit expression can be omitted, which is manifestly frame dependent. This equality will be evaluated in the Breit frame of the nucleons, $\boldsymbol{P}=0$. The reasons for this choice are known [4, 5] and will not be discussed here. Having chosen the Breit frame, we will express every frame-dependent quantity through an invariant quantity, which reduces to the starting one in the Breit frame. As a result, we are led after some straightforward algebra to the following frame independent equality:

$$\begin{aligned} i\left[\varepsilon_{\alpha 3\gamma}\langle p_2|A_\mu^\gamma|\,p_1\rangle + \frac{q'_\mu}{i}\,\langle p_2|S_{AV}^{\alpha 3}|\,p_1\rangle\right]& \\ = \frac{1}{\pi}\int_{-\infty}^{+\infty} \frac{\mathrm{d}\nu}{P^2}\,\tilde{\varrho}_\mu^\alpha(\nu, q^2(\nu), \kappa^2(\nu), t)\,.& \end{aligned} \tag{III.3}$$

where

$$P=\tfrac{1}{2}(p_1+p_2);\quad t=(p_2-p_1)^2\equiv\Delta^2;\quad \nu=P\cdot\kappa;\quad q'=q-(\nu/P^2)\,P;$$
$$q^2(\nu)=\nu^2/P^2+q'^2;\quad \kappa^2(\nu)=q^2+t+2q'\cdot\Delta;\quad P^2=m^2-t/4. \tag{III.4}$$

and

$$\tilde{\varrho}^{\alpha}_{\mu}=\frac{(2\pi)^4}{2}\sum\langle p_2|P\cdot A^{\alpha}|n\rangle\,\langle n|V^{\text{e.m.}}_{\mu}|p_1\rangle\,\delta(p_2+q-p_n)-\text{c.t.} \tag{III.5}$$

In Eq. (III.4) $q'^2, q'\cdot\Delta$ are two arbitrary parameters which "label" the sum rule Eq. (III.3). By the standard procedure, $\tilde{\varrho}^{\alpha}_{\mu}$ is now decomposed into three pieces:

$$\tilde{\varrho}^{\alpha}_{\mu}=\tilde{\varrho}^{\alpha\text{I}}_{\mu}+\tilde{\varrho}^{\alpha\text{II}}_{\mu}+\tilde{\varrho}^{\alpha\text{III}}_{\mu} \tag{III.6}$$

whose explicit expression can be found elsewhere [9].

Moreover, we write:

$$\varrho_{\mu}=u(\boldsymbol{p}_2)\,\gamma_5\,\{P_{\mu}P_1+\hat{q}'P_{\mu}P_2+\gamma_{\mu}P_3$$
$$+[\hat{q}'\gamma_{\mu}-q'_{\mu}]\,P_4+\Delta_{\mu}P_5+\hat{q}'\Delta_{\mu}P_6+q'_{\mu}P_7+\hat{q}'q'_{\mu}P_8\}\,u(\boldsymbol{p}_1) \tag{III.7}$$

where

$$P_i\equiv P_i(\nu,\,q^2(\nu),\,\kappa^2(\nu),\,t). \tag{III.8}$$

Eq. (III.3) can thus be decomposed into eight non trivial equations. We shall be interested only in relations between the real parts of the l.h.s. and r.h.s. Due to the crossing property of $\tilde{\varrho}_{\mu}$, they involve $\text{Re}\,P^{(0),(+)}_{1,2,4}$ and $\text{Re}\,P^{(-)}_{3,5,6,7,8}$. Now, one can easily see from the expression of $\varrho^{\text{II}}_{\mu}$ [9] that the contribution from the intermediate one pion state to the sum rule for $\text{Re}\,P_i$ contains just the real part of the corresponding physical amplitude P_i (Eq. II.6) in the point

$$\left(\nu=\sqrt{P^2(\mu^2-q'^2)};\quad \kappa^2=\mu^2+t+2q'\cdot\Delta;\quad t\right). \tag{III.9}$$

Varying q'^2 and $q'\cdot\Delta$ (which are the arbitrary parameters at disposal) the kinematical configuration varies. In particular, the physical (c.m.s.) threshold corresponds to the values:

$$q'^2=\frac{\mu^2}{4m(m+\mu)}(\kappa^2-\mu^2);\quad q'\cdot\Delta=\frac{\mu}{2(m+\mu)}(\kappa^2-\mu^2).$$

In this way, Eq. (III.3) supplies e.g. an expression for the physical amplitudes at threshold $\overline{P}_i(\text{th.})$, in terms of the e.t.c. in the l.h.s. and of the remaining contributions. Among these, the one coming from the one nucleon state in $\tilde{\varrho}^{\text{I}}_{\mu}$ is easily computed. The remaining ones will be generically called "corrections"; thus, the final aspect of the sum rules will be:

$$\overline{P}_i(\text{th.}) = \text{e.t.c.} + \text{Nucleon} + \text{“corrections”}. \tag{III.10}$$

An evaluation of these “corrections” whose typical order of magnitude is about 10% has been given [5, 9] and we do not discuss it here. Rather, we notice that we have obtained at this point predictions for five $(-)$ amplitudes and three $(+)$ amplitudes; therefore, more some rules are needed to derive the complete amplitude T_μ. This problem is fully discussed in Ref. [9], where it is shown that this can be done by means of “reasonable assumptions” about the commutator $[D_\alpha, V]$ (which supplies the remaining amplitudes) not necessarily requiring the use of “field algebra” hypotheses. Moreover, it is also shown in Ref. [9] that the problems related to the unknown Schwinger term can be completely avoided by substituting the amplitude $\overline{P}_7^{(-)}$, to which it contributes, with amplitudes related to $\overline{P}_7^{(-)}$ by the gauge condition and supplied by the commutator Eq. (IV. 2). As a conclusion, recalling Eqs. (II.17–23), the following expressions are found:

$$E_{0^+}^{(-)}(\text{th.}) = -\mathcal{R}\frac{G_A}{2f_\pi}(\kappa^2-\mu^2) + \text{“corrections”};$$

$$\mathcal{R} \equiv \sqrt{\frac{[2m+\mu)^2-\kappa^2]}{4m(m+\mu)}} \cdot [(2m+\mu)^2-\kappa^2]. \tag{III.11}$$

$$E_{0^+}^{(0)}(\text{th.}) = \mathcal{R}\frac{-\mu}{2m}\frac{G_A(0)}{2f_\pi}F_1^S(\kappa^2-\mu^2) + \text{“corrections”};\ f_\pi \simeq 0{,}67\,\mu.$$

$$\mathcal{L}_{0^+}^{(-)}/\kappa_0(\text{th.}) = \mathcal{R}\frac{D(\kappa^2-\mu^2)}{4m\mu f_\pi} + \text{“corrections”}.$$

$$\mathcal{L}_{0^+}^{(0)}/\kappa_0(\text{th.}) = \mathcal{R}\, G_A(0)\, F_1^S(\kappa^2-\mu^2)/4m f_\pi + \text{“corrections”}.$$

$$y^{(-)}/|\boldsymbol{\kappa}|\,|\boldsymbol{q}|\,(\text{th.}) = G_A(0) G_M^V(\kappa^2-\mu^2)\,(m+\mu)/4m^2\mu f_\pi + \text{“corrections”}.$$

$$y^{(0)}/|\boldsymbol{\kappa}|\,|\boldsymbol{q}|\,(\text{th.}) = \mu G_A(0)\, F_1^S(\kappa^2-\mu^2)/16\,m^3 f_\pi + \text{“corrections”}.$$

$$z^{(-)}/|\boldsymbol{\kappa}|\,|\boldsymbol{q}|\,(\text{th.}) = G_P(\kappa^2-\mu^2)/6m f_\pi - G_A(0)\, G_M^V(\kappa^2-\mu^2) \cdot (m+\mu)/12\,m^2\mu f_\pi + \text{“corrections”}.$$

$$z^{(0)}/|\boldsymbol{\kappa}|\,|\boldsymbol{q}|\,(\text{th.}) = G_A(0)\, G_M^S(\kappa^2-\mu^2)/12\,m\mu f_\pi + \text{“corrections”}.$$

$$v^{(-)}/|\boldsymbol{\kappa}|\,|\boldsymbol{q}|\,\kappa_0(\text{th.}) = G_P(\kappa^2-\mu^2)/2\mu^2 f_\pi(2m+\mu) + \text{“corrections”}.$$

$$v^{(0)}/|\boldsymbol{\kappa}|\,|\boldsymbol{q}|\,\kappa_0(\text{th.}) = -G_A(0)\, F_1^S(\kappa^2-\mu^2)/8\,m^2\mu f_\pi + \text{“corrections”}.$$

$$\chi^{(-)}/|\boldsymbol{q}|^2(\text{th.}) = -G_P(\kappa^2-\mu^2)\,(2m+\mu)^2/24m^2 f_\pi(m+\mu) + \text{“corrections”}.$$

$$\chi^{(0)}/|\boldsymbol{q}|^2(\text{th.}) = -G_A(0)\, F_1^S(\kappa^2-\mu^2)\,(2m+\mu)/24m^2\mu f_\pi + \text{“corrections”}.$$

Here G_A, G_P are the usual axial-vector form factor, i.e.

$$\langle p_2|A_\mu^\gamma|\, p_1\rangle = \bar{u}(p_2)(\tau^\gamma/2)\gamma_5[G_A\gamma_\mu + G_P\Delta_\mu]\, u(p_1);$$

$$G_A(0) = -1{,}18; \quad D(t) = -2m\, G_A(t) + t\, G_P(t).$$

Moreover, $F_1^{S,V}$ and $G_M^{S,V}$ are the usual nucleon electromagnetic form factors. The more elaborated derivation of the quantity F_1 (and the analogous longitudinal one F_5) can be found in Ref. [9].

IV. Comparison with the Experimental Data

Let us consider the photoproduction case. As far the S wave multipole is concerned, we see that it depends essentially from the fit we use for $G_A(t)$. Actually, we have used the "dipole fit"

$$G_A(t) = \frac{G_A(0)}{(1 - t/M_A^2)^2} \tag{IV.1}$$

and, fixing $G_A(0) = -1{,}18$, we have allowed M_A to assume the values 6, 7, 8, 9. Correspondingly, we obtain for the slope of the differential cross section:

$$\frac{|\boldsymbol{\kappa}|}{|\boldsymbol{q}|}\frac{d\sigma}{d\Omega}\bigg|^{(\pi^+)}_{(\text{th.})} = \begin{cases} 14{,}1\ \mu\text{b/ster}\,(M_A = 6\,\mu) \\ 14{,}5\ \mu\text{b/ster}\,(M_A = 7\,\mu) \\ 14{,}7\ \mu\text{b/ster}\,(M_A = 8\,\mu) \\ 14{,}9\ \mu\text{b/ster}\,(M_A = 9\,\mu). \end{cases} \tag{IV.2}$$

As a check of our solutions for the p-waves, we have evaluated the unpolarized differential cross section at 90% as a function of the incoming photon energy in the laboratory frame. Following the notation of Eq. (II.10–11) we give the result in Table 1.

Table 1. F_1, y, z, χ *eqns.* (2.11) *in* π^+ *photoproduction (units* $\mu = 1$*) as a function of the incoming photon energy in the laboratory frame* (MeV). A *is defined by eqn.* (2.10 *and* 11)

E_γ^L(MeV)	F_1	y	z	χ	F_1^2	$\frac{9}{2}z^2$	y^2	$3yz$	$3E_{0^+}\chi$	A
160	1.284	−0.185	0.229	−0.052	1.649	0.236	0.034	−0.127	−0.200	1.592
170	1,275	−0.273	0.337	−0.105	1.627	0.509	0.076	−0.276	−0.403	1.533
180	1.266	−0.353	0.436	−0.160	1.602	0.854	0.125	−0.462	−0.615	1.504
190	1.256	−0.437	0.536	−0.219	1.577	1.293	0.191	−0.703	−0.842	1.516
200	1.248	−0.518	0.638	−0.281	1.552	1.832	0.268	−0.991	−1.080	1.581
210	1.236	−0.594	0.732	−0.342	1.527	2.415	0.353	−1.304	−1.314	1.677
220	1.226	−0.667	0.824	−0.401	1,503	3.056	0.445	−1.649	−1.541	1.814

As it can be seen e.g. by looking at the data quoted by *P. Noelle*, *W. Pfeil* and *S. Schwela* [10], the agreement with the experimental data is good up to $E_\gamma^L \simeq 220$ MeV. To have another check, we have considered the asymmetry ratio, defined as

$$\Sigma(\vartheta) = \frac{\sigma_\perp - \sigma_\parallel}{\sigma_\perp + \sigma_\parallel}; \quad \sigma_{\perp,\parallel} = \sigma(1 \pm \Sigma). \tag{IV.3}$$

In terms of multipoles, this quantity reads, at $\vartheta = 90°$:

$$\Sigma(\vartheta = \pi/2) = -\frac{[9z^2 + 6E_{0^+} + \chi + 6yz]}{2A} + O(|\boldsymbol{q}|^4). \tag{IV.4}$$

Moreover we have also separately computed the quantities $d\sigma_\perp/d\Omega$, $d\sigma_L/d\Omega$ and the results are shown in Table 2.

Table 2. $d\sigma/d\omega^{(\mathrm{unpol})}$, $d\sigma_\perp/d\omega$ *asymmetry ratio* Σ *and* $\sigma_\parallel/\sigma_\perp$ *ratio at* $\theta = 90°$, *varying the photon laboratory energy*

E_γ^L(MeV)	(unpol) $\frac{d\sigma}{d\omega}$ (μb/ster) $(\theta=\pi/2)$	$\frac{d\sigma_\perp}{d\omega}$ (μb/ster) $(\theta=\pi/2)$	Σ	$\frac{d\sigma_\parallel}{dr_\perp}$
160	4.60	4.87	0.057	0.89
170	5.92	6.58	0.111	0.78
180	6.73	7.72	0.148	0.70
190	7.43	9.16	0.166	0.67
200	8.29	9.54	0.151	0.70
210	9.18	10.30	0.121	0.76
220	10.20	10.85	0.074	0.86

Again, the agreement with the experimental results [11] seems to be encouracing up to $E_\gamma^L \simeq 220$ MeV. It should be noticed that the numbers we give are based on the fit:

$$G_P(t) = \frac{2 f_\pi g_{\pi N}}{t - \mu^2}; \quad g_{\pi N} \simeq 13{,}5. \tag{IV.5}$$

which should work in the $\kappa^2 = 0$ case reasonably well.

Coming now to the electroproduction case, we shall consider only threshold results. A recent experiment of *Amaldi et al.* [12], corresponding to the set of values

$$\kappa^2 = -10{,}4\,\mu^2; \quad E_e^L = 800\ \text{MeV}; \quad E_e'^L = 542\ \text{MeV}; \quad \varepsilon = 0{,}74. \tag{IV.6}$$

gives the result

$$\lim_{|\boldsymbol{q}|\to 0} \frac{1}{|\boldsymbol{q}|} \frac{d^2\sigma}{d\Omega_e'^L dE_e'^L} = (4{,}9 \pm 14\%)\, 10^{-31} \frac{\text{cm}^2}{(\text{Gev}/c)^2}. \tag{IV.7}$$

Our result, for the values $M_A = 6, 7, 8, 9\,\mu$ is (in the same units)

$$3{,}83\,(6\,\mu);\quad 4{,}73\,(7\,\mu);\quad 5{,}50\,(8\,\mu);\quad 6{,}14\,(9\,\mu) \tag{IV.8}$$

and the ratio σ_L/σ_T:

$$0{,}65\,(6\,\mu);\quad 0{,}48\,(7\,\mu);\quad 0{,}39\,(8\,\mu);\quad 0{,}32\,(9\,\mu). \tag{IV.9}$$

Another low-energy measurement has been recently performed at Daresbury [13]. The extrapolation at threshold agrees with our theoretical predictions for a value of M_A close to 8; this is the value which could fit both experiments well. On the contrary, the value $M_A = 6\,\mu$ seems to be ruled out.

References

1. *Chew, G.F., Goldberger, M.L., Low, F., Nambu, Y.*: Phys. Rev. **106**, 1345 (1957).
2. *Fubini, S., Nambu, Y., Wataghin, V.*: Phys. Rev. **111**, 329 (1958).
3. *Berends, T.A., Donnachie, A., Weaver, D.L.*: Nucl. Phys. B **4**, 1 (1967). For an exhaustive list of references see also: *Adler, S.L.*: Ann. Phys. **50**, 189 (1968). *Gehlen, G. v.*: Threshold pion electro- and photoproduction. Bonn Univ. P1 2–80 (1970).
4. *Fubini, S., Furlan, G.*: Ann. Phys. **48**, 322 (1968).
5. *Furlan, G., Paver, N., Verzegnassi, C.*: Nuovo Cimento **62** A, 519 (1969); **70** A, 247 (1970).
6. A complete review about the subject can be found in Chapters 2 and 3 of the book "Current algebras and applications to particle physics" by *Adler, S.L., Dashen, R.F.*
7. For a complete list and discussion of these works in the photoproduction case, see: *De Baenst, P.*: An improvement on the Kroll-Ruderman theorem. To be published in Nucl. Phys. In the electroproduction case, see for instance: *Nambu, Y., Yoshimura, M.*: Phys. Rev. Letters **24**, 25 (1970). *Gleeson, A.M., Gundzik, M.G., Kuriyan, J.G.*: Phys. Rev. **173**, 1708 (1968).
8. *Gross, D.J., Jackiw, R.*: Phys. Rev. **163**, 1688 (1967).
9. *Verzegnassi, C.*: Low energy pion photo- and electroproduction on nucleons by current algebra" DNPL/R 8.
10. *Noelle, P., Pfeil, W., Schwela, D.*: Multipole analysis of pi-zero and pi-plus photoproduction between threshold and 500 MeV. Bonn Univ. PI 2–79 (1970).
11. *Grilli, M., Nigro, M., Schiavuta, E., Soso, F., Spillantini, P., Valente, V.*: Nota Interna n. 356, LNF-67/18 (1967).
12. *Amaldi, E., Borgia, B., Pistilli, P., Bolla, M., Di Giorgio, G.V., Giazotto, A., Serbassi, S., Stoppini, G.*: Nuovo Cimento **65** A, 377 (1970).
13. *Nuthakki, N.V.P.R., Turner, O.T., Dickinson, B., Ibbotson, M., Shuttleworth, J., Siddle, R., Sofair, A., Fannon, J., Allison, J., Clegg, J., Foster, F., Hellings, R.D.*: Daresbury Preprint DNLP/P 50.

Dr. *C. Verzegnassi*
Istituto di Fisica Teorica
Università degli Studi di Trieste
Trieste/Italy

Pion Electroproduction in the Low-Energy Region

G. von Gehlen

Contents

I. Kinematics

I.1. Introduction

In the past few years electroproduction experiments at high energies and high momentum transfers have given unexpected and exciting results. However, in the low-energy region there has been considerable activity too. Many experiments on single pion electroproduction in the resonance region are presently being performed [1–6, 46]. In the following talk I want to summarize what can be learned from these low-energy experiments. I also want to talk about some recent theoretical work which has been done in order to interpret these experiments.

Our definition of the kinematic variables is shown in Fig. 1. The masses of the nucleon, pion and electron are called m, μ and m_e, respectively. As is usual, we assume the electroproduction to go by a one photon exchange mechanism. Since the electron-photon interaction is explicitly known, the interesting information obtained from electro-

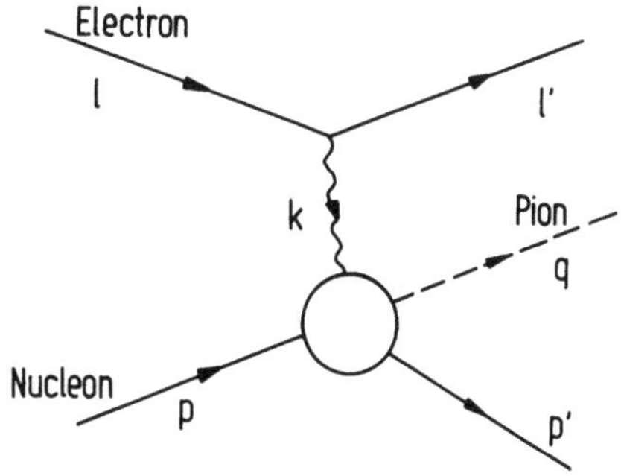

Fig. 1. Kinematics of single pion electroproduction

production is on the virtual photoproduction process represented by the lower bubble of Fig. 1.

Apart from polarizations, the virtual photoproduction amplitude depends on the three independent variables

$$s = -(p+k)^2; \quad t = -(k-q)^2 \tag{1.1}$$

and the photon momentum transfer

$$\lambda^2 = \boldsymbol{k}^2 - k_0^2 . \tag{1.2}$$

Instead of s we also use W with $s = W^2$ and the quantity $E_\gamma = (s - m^2)/(2m)$. E_γ is the lab energy which a real photon in photoproduction must have in order to produce the final pion-nucleon c.m.-energy W.

Three different frames of reference are useful for the evaluation of three-momenta, energies, polarizations etc.:

1. The laboratory frame $\boldsymbol{p}_\mathrm{L} = 0$ (this will be indicated by a subscript "L").

2. The Breit frame of the scattered electron $\boldsymbol{l}_\mathrm{B} = -\boldsymbol{l}'_\mathrm{B}$ or $k_{0\mathrm{B}} = 0$ (denoted by a subscript "B").

3. The center-of-mass frame of the virtual photoproduction $\boldsymbol{p}' + \boldsymbol{q} = 0$ (three momenta etc. written without subscript).

The relation between the photon momentum transfer and the electron scattering angle α is most simply expressed in the lab frame:

$$\lambda^2 = 4\,E E' \sin^2(\alpha_\mathrm{L}/2) \tag{1.3}$$

where we have used E' and E for the lab electron energies.

The c.m.-system of the final hadrons is useful for the study of the final state interaction between the final pion and nucleon. For this purpose one performs a multipole decomposition of the virtual photoproduction amplitude in the c.m. and makes use of Watson's theorem.

I.2. The Polarization of the Virtual Photon

The Breit frame is most convenient for the description of the polarization properties of the virtual photon. Generally, in the Breit frame a relativistic electron is coupled only to transversely polarized photons:

Introducing the operators

$$a_{\pm} = (1 \pm \gamma_5)/2 \tag{1.4}$$

which for relativistic electrons project pure helicity $\pm 1/2$-states from the electron spin function u, the electron-photon vertex can be decomposed as follows

$$\begin{aligned}\bar{u}\gamma_\mu u &= \bar{u}(a_+ + a_-)\gamma_\mu(a_+ + a_-)u \\ &= \bar{u}a_+\gamma_\mu a_- u + \bar{u}a_-\gamma_\mu a_+ u\end{aligned} \tag{1.5}$$

because

$$a_+\gamma_\mu a_+ = a_-\gamma_\mu a_- = 0. \tag{1.6}$$

The spin orientations corresponding to the two terms in the last line of Eq. (5) are depicted in Fig. 2, where we see that the photon must have helicity $+1$ and -1, respectively. Thus in the Breit frame unpolarized electrons will give rise to an equal mixture of photons with helicity ± 1 and no population of the helicity zero state. By a Lorentz

Fig. 2. The electron-photon interaction as it appears in the Breit frame. The helicities of the electron and photon are indicated by double arrows

transformation from the Breit frame to the hadron c.m. the photon acquires longitudinal components and the transverse part is no longer an equal mixture of helicity $+1$ and -1 but now shows the transverse polarization ε,

$$\varepsilon = \left(1 + 2\tan^2\frac{\alpha_B}{2}\right)^{-1} = \left(1 + \frac{2W^2k^2}{m^2\lambda^2}\tan^2\frac{\alpha_L}{2}\right)^{-1}. \tag{1.7}$$

Since the photon pole of the graph Fig. 1 forces the electron to be scattered preferrably in the forward direction, the cross sections are largest when α_L is small, i.e. it is easier to produce a large polarization ε than to produce virtual photons with ε small.

I.3. The Form of the Cross Section

We now consider the structure of the cross section for a few not too complicated experimental arrangements.

1. If only the final electron is detected ("single arm experiment"), 2 structure functions σ_t and σ_l, each depending on the 2 variables s and λ^2, can be measured [7]:

$$\frac{d\sigma}{d\Omega_e^L \, dE'} = \Gamma_t(\sigma_t + \varepsilon\sigma_l). \tag{1.8}$$

Here $d\Omega_e^L$ is the solid angle of the final electron measured in the lab frame, and

$$\Gamma_t = \frac{1}{2\pi^2 137} \frac{E'}{E} \frac{E_\gamma}{\lambda^2} \frac{1}{1-\varepsilon} \tag{1.9}$$

is a measure of the flux of virtual photons emitted by the scattered electron. σ_t and σ_l are linearly related to the structure functions W_1 and W_2 which are known from deep inelastic electron scattering. In fact (see e.g. Ref. [8]):

$$W_1 \sim \sigma_t \quad \text{and} \quad W_2 \sim \sigma_t + \sigma_l. \tag{1.10}$$

2. If a final hadron is detected in coincidence with the scattered electron, 4 structure functions A, B, C and D, each depending on the 3 variables s, t and λ^2 (or, equivalently, W, θ, λ^2) can be measured:

$$\frac{d\sigma}{d\Omega_e^L \, dE' \, d\Omega_\pi} = \Gamma_t \frac{d\sigma_v}{d\Omega_\pi} \tag{1.11}$$

where the differential virtual photoproduction cross section is [9, 10].

$$\frac{d\sigma_v}{d\Omega_\pi} = A + \varepsilon B + \varepsilon \sin^2\theta \cos 2\varphi \, C + \sqrt{\varepsilon(1+\varepsilon)} \sin\theta \cos\varphi \, D. \tag{1.12}$$

In Eq. (12), θ is the production angle of the pion in the hadron c.m.-system, φ is the azimuth of the orientation of the electron scattering plane relative to the final pion-nucleon plane, measured around the photon direction in the Breit frame. Observe that for convenience $d\Omega_e^L$ is taken in the lab frame whereas $d\Omega_\pi$ is measured in the virtual photoproduction c.m. frame.

A and C arise from purely transversely polarized photons, B is due to purely longitudinal photons. D contains the interference between longitudinal and transverse contributions. In the limit $\lambda^2 \to 0$, B and D vanish and $d\sigma_v/d\Omega_\pi$ goes over into the well-known photoproduction cross section for linearly polarized photons with polarization ε:

$$\frac{d\sigma}{d\Omega_\pi} = A(W, \theta) + \varepsilon \sin^2\theta \cos 2\varphi \, C(W, \theta). \tag{1.13}$$

Comparing Eqs. (8) and (11, 12) we see that

$$\sigma_t(W,\lambda^2)=\int d\Omega_\pi A(W,\theta,\lambda^2); \qquad \sigma_l(W,\lambda^2)=\int d\Omega_\pi B(W,\theta,\lambda^2). \tag{1.14}$$

3. More detailed information on the electroproduction matrix elements can only be obtained if polarized incident particles are used or if the polarization of a final particle is observed. Consider a single arm experiment on a polarized nucleon target with polarization $|\boldsymbol{P}|$ in the direction $\boldsymbol{P}$. One finds that the cross section is of the same form as in Eq. (8) but with one additional term containing a new structure function $\sigma_P(W,\lambda^2)$:

$$\frac{d\sigma}{d\Omega_e^L\, dE'}=\Gamma_t(\sigma_t+\varepsilon\sigma_l+\boldsymbol{P}(\boldsymbol{l}\times\boldsymbol{l}')\sqrt{\varepsilon(1+\varepsilon)}\,\sigma_P). \tag{1.15}$$

Christ and *Lee* [11] have shown that σ_P must vanish if time reversal invariance is valid. So the detection of a non-zero asymmetry of the distribution of the scattered electron with respect to the plane orthogonal to the nucleon polarization would be a clear test of T-invariance violation. The only assumption is the validity of the single photon exchange mechanism.

4. An effect of the polarization of the incident electron can only be seen in a coincidence experiment. Only the longitudinal component of the electron polarization contributes to an asymmetry effect. If the degree of the longitudinal electron polarization is ζ, the cross section is found to be [12]

$$\frac{d\sigma}{d\Omega_e^L\, dE'\, d\Omega_\pi}=\Gamma_t\left\{\left(\frac{d\sigma_v}{d\Omega_\pi}\right)_{\text{unpolar.}}+\zeta\sqrt{\varepsilon(1-\varepsilon)}\sin\theta\sin\varphi\, F(W,\theta,\lambda^2)\right\} \tag{1.16}$$

where the unpolarized coincidence cross section is given as in Eqs. (11) and (12).

Because of the factor $\sqrt{\varepsilon(1-\varepsilon)}$ the polarization dependent term produces an asymmetry which is strongest at medium angles α_L and, because of the factor $\sin\varphi$, it is orthogonal to the asymmetry created by the transverse-longitudinal interference term D.

Since the electroproduction amplitude depends on six complex invariant function of s, t and λ^2 [13, 14] it is not possible to obtain the amplitudes from experiment without recourse to theoretical ideas if not several other more complicated polarization experiments are performed. Cross section formulae for more involved experimental arrangements (i.e. coincidence experiments on a polarized nucleon target [15, 47]) have been worked out, but we shall not give details here because these experiments will be very difficult.

Experiments of type 1, 2 and 3 have been performed. Examples of the results will be shown below in Figs. 3, 4, 7, 10 and 12.

II. Interesting Aspects of Low-Energy Electroproduction Experiments

II.1. Tests of one Photon Exchange, the Ratio σ_l/σ_t

The fundamental assumption of a pure one photon exchange mechanism allowed us to consider electroproduction experiments to be essentially virtual photoproduction experiments. Since all the decompositions of the cross section formulae of the last section are based on the one photon exchange assumption, this assumption may be tested by checking the ε and φ-dependence of the coincidence cross section in (1.12) or, more simply and less stringent by testing the linear dependence of the single arm cross section, Eq. (1.8), on ε.

The test of the linear dependence of Eq. (1.8) on ε is a generalization of the famous Rosenbluth test which is familiar from elastic electron scattering. In fact, the Rosenbluth formula may be written in the form (1.8) with [8]

$$\sigma_t(W, \lambda^2) = \frac{4\pi^2\alpha}{E_\gamma} \frac{\lambda^2}{4m^2} G_m^2(\lambda^2)\, \delta\left(E' - E + \frac{\lambda^2}{2m}\right) \tag{2.1}$$

and

$$\sigma_l(W, \lambda^2) = \frac{4\pi^2\alpha}{E_\gamma} G_e^2(\lambda^2)\, \delta\left(E' - E + \frac{\lambda^2}{2m}\right). \tag{2.2}$$

Here G_e and G_m are the Sachs form factors of the nucleon.

Fig. 3 shows a plot of experimental values of

$$\Gamma_t^{-1} \frac{d\sigma}{d\Omega_e^L\, dE'} = \sigma_t + \varepsilon\sigma_l \tag{2.3}$$

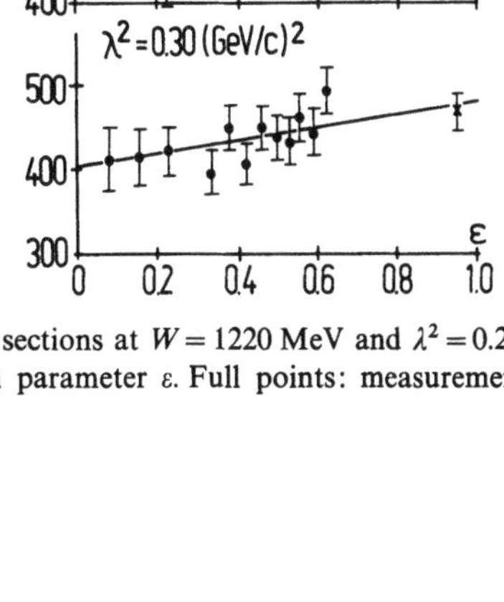

Fig. 3. Single arm cross sections at $W = 1220$ MeV and $\lambda^2 = 0.2$ and 0.3 $(GeV/c)^2$ plotted against the polarization parameter ε. Full points: measurements of Ref. [41], crosses: Ref. [16]

against ε for $\lambda^2 = 0.2$ and 0.3 $(\mathrm{GeV}/c)^2$. Since the error bars are still quite large and furthermore, since the experimental points at small and large values of ε were not obtained in the same experiment (so that systematic errors may be present) the test is not yet very conclusive but certainly compatible with a straight line*. However, such a plot is also useful for reading off the values of σ_t and σ_l. From Fig. 3 we see that $\sigma_l \ll \sigma_t$ for the W and λ^2 considered. In the region of the first resonance σ_l is found to reach a maximum around $\lambda^2 \approx 0.3 \ldots 0.4$ $(\mathrm{GeV}/c)^2$, falling off strongly for higher values of λ^2, see Fig. 4. At $\lambda^2 = 0$ σ_l vanishes kinematically.

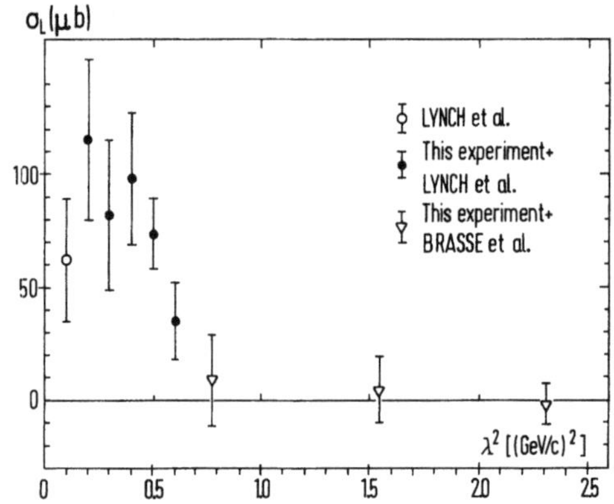

Fig. 4. Longitudinal virtual photoproduction cross section σ_l at the energy of the first resonance plotted as a function of λ^2 (from *Bartel et al.* [16])

II.2. The Form Factor of the Vertex $(\gamma N \Delta)$

In a simple theory the virtual photoproduction at low energies may be described by the four graphs of Fig. 5 (eventually plus graphs with higher resonances in the s-channel, resonances in the u-channel and ϱ- and ω-exchange in the t-channel). Consider the graph in Fig. 5d which through the vertex on the left hand side depends on the transition form factor $(\gamma N \Delta)$. If it were possible to isolate the contribution of graph Fig. 5d from the whole electroproduction matrix element and if a complete measurement of the matrix element were available, we could deduce the $(\gamma N \Delta)$-vertex form factor from experiment. Obviously both, the isolation of the graph and a complete measurement are complicated problems. The mere addition of graphs may in fact lead to double counting. However, a complete measurement may provisionally be substituted by a reasonably consistent theoretical model which describes

* *Note added in proof:* For a new experiment, see [49].

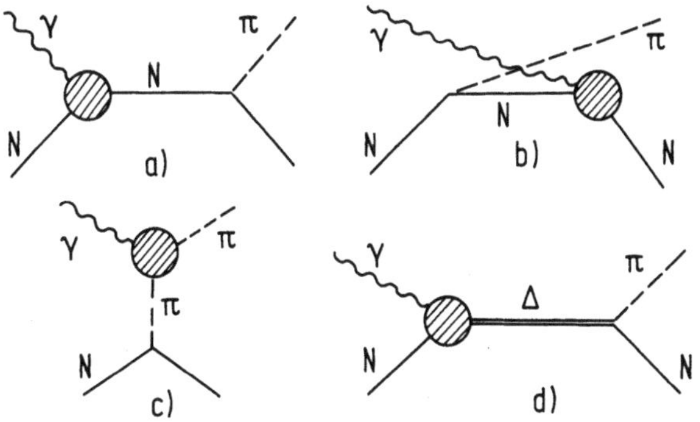

Fig. 5a–d. The four most important diagrams for a simple low-energy electroproduction theory. The bubbles indicate those vertices in which interesting electromagnetic form factors are involved

the existing experimental data, e.g. the model which will be described in Sec. III.3 below.

The form factors for the vertex $(\gamma N\Delta)$ in the isobar graph Fig. 5d can be related to the multipoles M_{1+}, E_{1+} and L_{1+} (obtained from a detailed experiment or from a theoretical model) as follows:

We define the vertex $(\gamma N\Delta)$ by the expression [17, 18]

$$\bar{u}_\mu(\boldsymbol{Q})\, V_\mu u(\boldsymbol{p}) \tag{2.4}$$

with

$$V_\mu = \left(C_3(\lambda^2)\gamma_\nu - \frac{i}{\mu} C_4(\lambda^2)\, p_\nu - \frac{i}{\mu} C_5(\lambda^2)\, Q_\nu \right) \cdot \gamma_5 (k_\nu \varepsilon_\mu - k_\mu \varepsilon_\nu) . \tag{2.5}$$

Here C_3, C_4 and C_5 are the three form factors of the vertex $(\gamma N\Delta)$. Q, p and k are the momenta of the Δ, the nucleon and the photon, respectively, ε_μ is the photon polarization vector and $Q = p + k$. With this vertex, the graph of Fig. 5d leads to the amplitude

$$T = -\frac{e_r \lambda_1 \mu^{-2}}{s - m^{*2} + im\Gamma} \bar{u}(\boldsymbol{p}')q_\mu \pi_{\mu\nu} V_\nu u(\boldsymbol{p}) . \tag{2.6}$$

$\pi_{\mu\nu}$ is the numerator of the spin-$\frac{3}{2}$-propagator:

$$\pi_{\mu\nu} = (-i(\gamma Q) + m^*) \left(\delta_{\mu\nu} - \frac{\gamma_\mu \gamma_\nu}{3} + \frac{i}{3m^*}(\gamma_\mu Q_\nu - \gamma_\nu Q_\mu) + \frac{2}{3m^{*2}} Q_\mu Q_\nu \right) . \tag{2.7}$$

m^* is the Δ-mass, Γ its width and λ_1 is the $\pi N\Delta$-coupling constant. Eventual terms proportional to $s - m^{*2}$ have been dropped in Eq. (7) because a T-product is not unique off mass-shell.

In order to express the amplitude T in terms of multipoles, we evaluate Eq. (6) in the system $\boldsymbol{Q}=0$. In the calculation we put $Q_0=m^*$ in order to obtain simple formulae and because through $\pi_{\mu\nu}$ there is an arbitrariness in the definition of Eq. (6) for $s \neq m^{*2}$ anyway. In the system $\boldsymbol{Q}=0$, T is simply related to the usual virtual photoproduction c.m. amplitude $\mathscr{F}$ (which is sandwiched between Pauli spinors χ):

$$T=\frac{4\pi W}{m}\chi^+ \mathscr{F}\chi. \tag{2.8}$$

$\mathscr{F}$ has a simple expansion in terms of multipoles [14] and we find [19]

$$\begin{gathered} C_3=\frac{M_{1+}+E_{1+}}{(W+m-k_0)\,f}; \quad C_4=-\frac{2\mu(E_{1+}-L_{1+})}{k^2 f}; \\ \frac{\mu C_3}{W}+\left(1+\frac{\lambda^2}{W k_0}\right)C_4+C_5=-\frac{2\mu E_{1+}}{W k_0 f} \end{gathered} \tag{2.9}$$

with

$$f=-\frac{e_r\lambda_1 q k}{12\pi\mu^2}\sqrt{\frac{W+m-q_0}{W+m-k_0}}\,\frac{1}{s-m^{*2}+im\Gamma}. \tag{2.10}$$

The width Γ does not need to be constant, it may depend on W.

On the other hand, using a completely analogous isobar model for pion-nucleon scattering, we relate λ_1 and Γ to the $\pi-N$-phase shift α_{33}:

$$\frac{e^{i\alpha_{33}}\sin\alpha_{33}}{q^3}=-\left(\frac{\lambda_1}{\mu}\right)^2\frac{W+m-q_0}{12\pi}\,\frac{1}{s-m^{*2}+im\Gamma} \tag{2.11}$$

so that

$$f=\frac{e_r\,k\,\sin\alpha_{33}\,e^{i\alpha_{33}}}{\lambda_1 q^2\sqrt{(W+m-k_0)(W+m-q_0)}} \tag{2.12}$$

independent of the energy dependence of Γ as long as we assume that the same Γ can be used in $\pi-N$-scattering and in electroproduction. From the Watson theorem we know that the multiples M_{1+}, E_{1+} and L_{1+} appearing in Eqs. (9) have the phase α_{33}. So, inserting (12) into Eqs. (9), we are able to express C_3, C_4 and C_5 solely in terms of the absolute values of M_{1+}, E_{1+} and L_{1+} (for isospin $I=3/2$) and the experimental pion-nucleon phase shift α_{33}. λ_1 has to be determined from Eq. (11), it is fixed by the value of Γ at $W=m^*$.

If the description of M_{1+}, E_{1+} and L_{1+} in terms of only the isobar graph makes sense at least near $W=m^*$, then C_3, C_4 and C_5 as determined from Eqs. (9) and (12) should turn out to be practically energy independent in the resonance region. Fig. 6 shows numerical results for the C_i at $\lambda^2=0.6$ (GeV/c)^2 as obtained using the resonant multipoles of the theoretical model of Ref. [20]. One sees that only C_3 is found to be

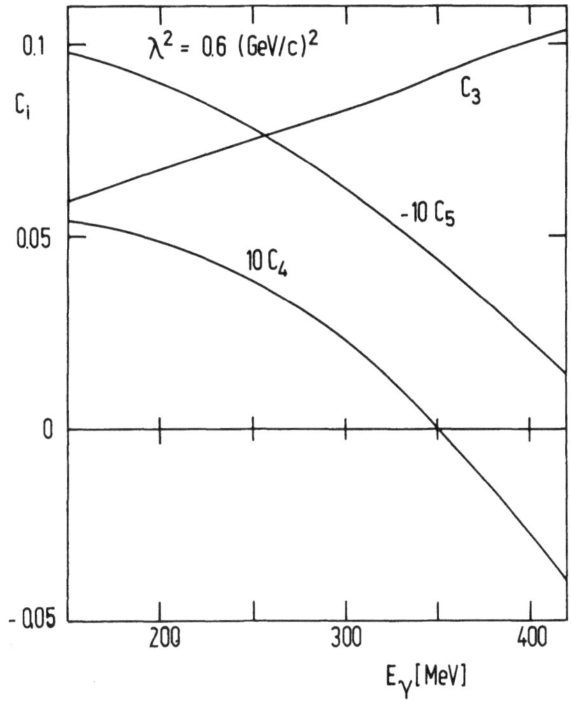

Fig. 6. The three $(\gamma N \Delta)$-transition form factors C_3, C_4 and C_5, defined in Eq. (2.5), as obtained from the dispersion calculation of Ref. [20] for $\lambda^2 = 0.6\,(\mathrm{GeV}/c)^2$. (The angular distribution curves shown in Fig. 12 below – full curves – are obtained from the same calculation.) The normalisation is such that $C_3(\lambda^2=0) \approx 0.35$ or $C_3(\lambda^2=0) \cdot m/\mu \approx 2.4$ (see Ref. [18]). For the $(\pi N \Delta)$-coupling we use $\lambda_1 = 2.07$

reasonably energy independent which of course is due to the nice resonance structure of M_{1+}. In order to obtain on energy independent value for C_4 and C_5 one must not use the whole multipoles E_{1+} and L_{1+} (always for $I = 3/2$) in Eq. (9) but some nonresonant background must be subtracted and be ascribed to other graphs than Fig. 5d. A practical way of performing such a subtraction would be possible from the Argand diagrams of E_{1+} and L_{1+}.

The definition (5) has the disadvantage that the big multipole M_{1+} appears both in C_3 and C_5, so that one is not able to recover easily the small multipoles E_{1+} and L_{1+} from the C_i. Other definitions have been given, see e.g. Ref. [21].

II.3. Form Factors of Higher Resonances

If the single-arm cross section (1.8) shows a strong resonance structure, one may simply define the background by drawing a smooth line below

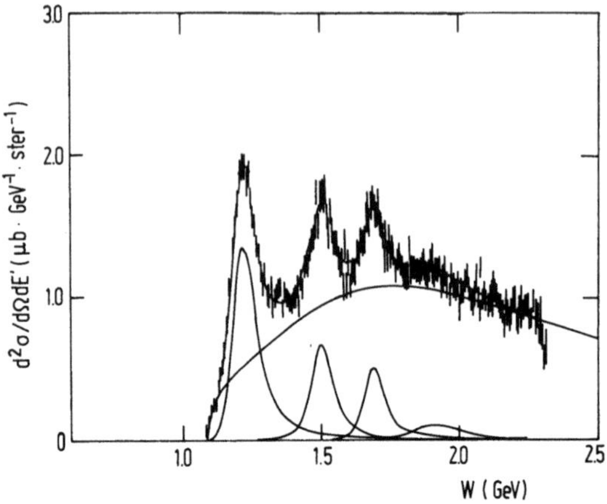

Fig. 7. Excitation curve at fixed angle $\alpha_L = 6°$ and fixed incident electron energy $E = 10$ GeV showing the electroexcitation of the first four nucleon resonances (from *Bloom et al.* [22]). The smooth line indicates the background obtained by subtracting the Breit-Wigner resonance contributions which are shown at the bottom

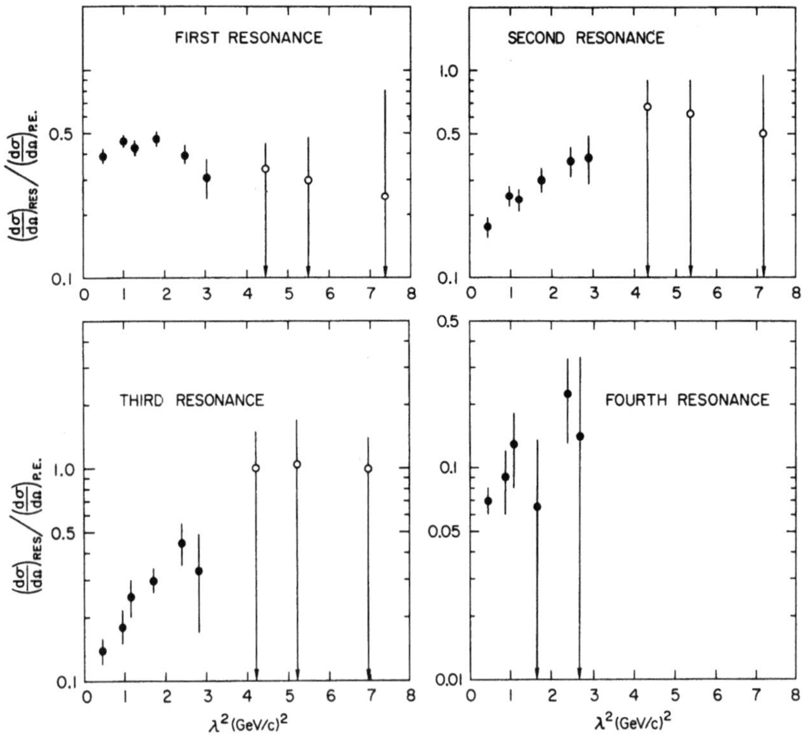

Fig. 8. Resonance cross sections integrated over E' and divided by the corresponding purely elastic (P.E.) scattering cross section, plotted against λ^2 (from Ref. [22]). The quantity plotted is proportional to the square of the resonance transition form factor divided by the square of the elastic form factor

the resonance peaks as it is done in Fig. 7 (see also Ref. [16]). One then relates the area between the smooth line and the measured curve to one effective form factor for the vertex ($\gamma N N^*$) without attempting to resolve the spin structure of the vertex. E.g. the authors of the recent SLAC experiment [22, 43] plotted the ratio of $\mathrm{d}\sigma/\mathrm{d}\Omega^{\mathrm{L}}_{e\mathrm{RES}}$ (area between resonance peak and smooth background curve) to $\mathrm{d}\sigma/\mathrm{d}\Omega^{\mathrm{L}}_{e}$ for the corresponding elastic scattering cross section. As Eqs. (1) and (2) show, this ratio is proportional to the square of the ratio of the resonance form factor to the elastic form factor. Results are shown in Fig. 8.

If experiments at different values of ε become available, effective longitudinal and transverse form factors may be obtained separately and may be compared to dispersion theory and quark model predictions.

II.4. The Electromagnetic Pion Form Factor

At present, only two methods seem to be feasible in order to obtain information on the electromagnetic form factor of the pion, F_π, in the spacelike region. One is to measure the Coulomb interference in pion-helium scattering. An experiment using this method has been carried out [23], however, so far the difficulties in handling the complicated nuclear physics corrections have prevented definite conclusions.

Very difficult at the moment is high-energy pion-electron scattering at reasonably high momentum transfer. This has to wait for the next generation of accelerators, but then it is expected to give useful results. For the present time it seems that the only practical way of getting experimental information on F_π in the spacelike region is to isolate the contribution of the pion exchange graph Fig. 5c in charged pion electroproduction.

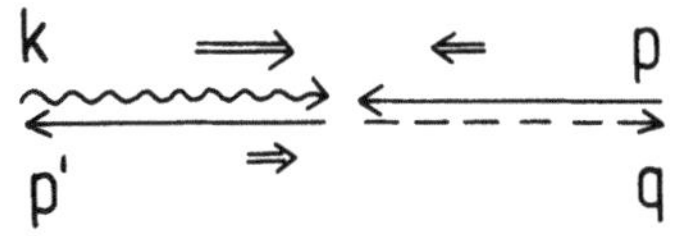

Fig. 9. Kinematical configuration of virtual photoproduction at $\theta=0°$. The helicities are indicated by double arrows, showing that for transverse photons a nucleon helicity flip is necessary.

At the unphysical point $t=\mu^2$ the contribution of the pion pole graph dominates completely. However, from helicity conservation it follows that at the physical point which is closest to $t=\mu^2$, that is at $\theta=0^0$, the cross section will vanish for *transverse* photons because a spin flip is needed, see Fig. 9. So, at low energies the contribution of the pion pole will be strongly suppressed in the transverse cross section. Indeed an explicit calculation shows that the transverse multipoles depend only very weakly on F_π for energies up to the first resonance [27].

For longitudinal photons (which have zero helicity and do not need to transfer their helicity to the nucleon) no spin flip factor is present near $\theta = 0^0$ and indeed one finds that the longitudinal cross section B of Eq. (1.12) has a forward peak due to the pion pole. So, in order to obtain information on F_π at low energies one must measure longitudinal cross sections.

Since an actual extrapolation to the pion pole is too difficult one usually assumes that $B(W, \theta, \lambda^2)$ near $\theta = 0^0$ is given by the Born approximation. But even with this assumption one does not get a direct relation between $B(W, \theta = 0^0, \lambda^2)$ and $F_\pi(\lambda^2)$ because the Born approximation depends also on the very poorly known electric neutron form factor $G_e^N(\lambda^2)$. The best way out would be to determine F_π and G_e^N simultaneously from $A(W, \theta, \lambda^2)$, $B(W, \theta, \lambda^2)$ and $D(W, \theta, \lambda^2)$, but, since the dependence of B on G_e^N is not very strong, one may also simply put $G_e^N = 0$ for not too large values of λ^2.

Up to now, two electroproduction coincidence experiments have been performed in order to get information on F_π. A Cornell experiment [10] measured at $\theta = 0^0$ and separated A and B (at $\theta = 0^0$ C and D do not contribute anyway, see Eq. (1.12)) by varying ε. In the CEA experiment [24], no variation of ε was performed, but instead angular distributions in φ for $0^0 < \theta < 50^0$ were measured. Then the information on F_π was extracted by fitting the data to the theoretical calculations of *Adler* [25] and *Zagury* [26]. The result is that F_π seems to fall off similarly to the nucleon form factors. However, since the highest value of λ^2 used was $\lambda^2 = 0.4\,(\mathrm{GeV}/c)^2$ and because of uncertainties in the theoretical models used and of experimental errors, as yet no distinction is possible between the behaviours

$$F_\pi \approx F_1^V; \qquad F_\pi \approx \frac{1}{1 + \lambda^2/m_\varrho^2}; \qquad F_\pi \approx G_e^P. \tag{2.13}$$

Here, F_1^V is the isovector Dirac nucleon form factor, the second case is ϱ-dominance and G_e^P is the Sachs proton form factor.

We remark that generally experiments with ε small are much more difficult than experiments with $\varepsilon \approx 0.9$ because for small ε the cross section is much smaller (see Eq. 1.9) and one needs high incident electron energies E in order that the final electron energy E' does not become too small for a given value of λ^2.

It is interesting to observe that π^--electroproduction should be more sensitive to F_π than the more conventional π^+-production [27]. This arises as follows:

Decomposing B into multipoles, we find (we use $\hat{S}_{l\pm} = \sqrt{\lambda^2} L_{l\pm}/k_0$):

$$B\left(\theta = \begin{matrix} 0^\circ \\ 180^\circ \end{matrix}\right) = \frac{Wq}{mE_\gamma} |\hat{S}_{0+} \pm \hat{S}_{1-} + 4\hat{S}_{1+} \pm 4\hat{S}_{2-} + \ldots|^2 \tag{2.14}$$

Now direct inspection of the formulae shows that in the Born approximation $\hat{S}^{\pi^+}_{0+}$ is strongly reduced by cancellations between different terms, whereas $\hat{S}^{\pi^-}_{0+}$ is not. So

$$\hat{S}^{\pi^-}_{0+} \gg \hat{S}^{\pi^+}_{0+} \tag{2.15}$$

and B^{π^-} should be appreciably larger than B^{π^+}, which entails that also σ_l/σ_t should be much larger for π^--production than for π^+-production.

It turns out that $\hat{S}_{1-}$ and $\hat{S}_{2-}$ are almost F_π-independent. Since furthermore $\hat{S}_{1+}$ is unconvenient for the determination of F_π because of its resonance enhancement, the determination of F_π at low energies can only rely on the S-wave multipoles $S^{\pi^\pm}_{0+}$ *.

II.5. The Electric Form Factor of the Neutron

The form factor $G_e^N(\lambda^2)$ enters the electroproduction matrix element through the Born graph Fig. 5b for π^+-production from protons. Both σ_t and σ_l turn out to be sensitive to G_e^N. So one may hope to obtain information on G_e^N more easily from electroproduction than from electron-deuteron scattering since in the electroproduction experiment no problems of the deuteron structure are involved [28]. However, there is no kinematically accessible region where the graph Fig. 5b is expected to dominate strongly and one is forced to study the G_e^N-dependence of the whole amplitude. It turns out [27] that for this purpose it is not sufficient to describe the amplitude in the Born approximation because the dispersion contributions are probably of the same order of magnitude as the variation expected for σ_t by varying G_e^N within reasonable limits. The relevant dispersion contributions cannot be calculated with sufficient accuracy at the moment, so, at present the prospects of getting interesting information on G_e^N from electroproduction do not look good.

II.6. The Axial Vector Nucleon Form Factor

Low-energy electroproduction can also give information on $G_A(\lambda^2)$, the axial vector form factor of the nucleon, if current algebra commutation relations and soft pion techniques are used. Since the application of current algebra to electroproduction will be discussed in detail in the talk of Dr. *Verzegnassi* [29], here we only mention that current algebra and PCAC lead to the relation (see e.g. Ref. [30]):

$$G_A(\lambda^2)/G_A(0) - F_1^V(\lambda^2) = \frac{2m\lambda^2}{e_r g_r} A_6^{(-)}\,(\text{soft pion}, \lambda^2) \tag{2.16}$$

---

* Three new experiments aiming to determine F_π have been published recently, see Refs. [50], [51], [52]. An experiment on pion-electron scattering is in progress at Dubna.

where F_1^{V} is the first Dirac isovector nucleon form factor, $e_r g_r = 4.11$ and $A_6^{(-)}$ is the odd isovector part of one of the conventional six relativistic electroproduction amplitudes. The term "soft pion" in the argument of A_6 means that the amplitude A_6 has to be evaluated at the unphysical point

$$s = u = m^2; \quad t = -\lambda^2; \quad \boldsymbol{q}^2 = q_0^2. \tag{2.17}$$

Since A_6 does not receive large contributions from the first resonance multipoles it is difficult to estimate the right hand side of Eq. (16). However, a dispersion calculation indicates [31] that

$$\frac{2m\lambda^2}{e_r g_r} A_6^{(-)} \,(\text{soft pion}, \lambda^2) \ll 1 \tag{2.18}$$

which means that we should have

$$G_{\mathrm{A}}(\lambda^2)/G_{\mathrm{A}}(0) \approx F_1^{\mathrm{V}}(\lambda^2), \tag{2.19}$$

in agreement with the results of neutrino experiments.

II.7. Test of Time-reversal Invariance

We conclude our enumeration of interesting aspects of low-energy electroproduction experiments with the clear test on T-invariance which has been proposed by *Christ* and *Lee* [11]. In Eq. (1.15) we already gave the form of the relevant cross section for a single arm experiment on a polarized nucleon target.

If we express σ_P in terms of multipoles, we see that σ_P is a transverse-longitudinal interference term:

$$\sigma_P = \frac{4\pi\sqrt{2}\,Wq}{mE_\gamma} \operatorname{Im}\{E_{0+}\hat{S}_{0+}^* - M_{1-}\hat{S}_{1-}^* + 2(M_{1+} + 3E_{1+})\hat{S}_{1+}^* - 2(3M_{2-} - E_{2-})\hat{S}_{2-}^* + \cdots\}. \tag{2.20}$$

If T-invariance is valid, it follows from the Watson theorem that in the elastic region each term $\operatorname{Im} E_{0+}\hat{S}_{0+}$, $\operatorname{Im} M_{1-}\hat{S}_{1-}$ etc. vanishes separately. A nonvanishing result for σ_P in the low-energy region would therefore prove a violation of the Watson theorem.

In Eq. (20) we omitted the charge indices. However, if the final hadrons are not observed, e.g. on a proton target one measures the sum of π^+- and π^0-production, which reads expressed in terms of the isoscalar amplitude A^0 and the two isovector amplitudes $A^{1/2}$ and $A^{3/2}$:

$$|A^{\gamma P \to n\pi^+}|^2 + |A^{\gamma P \to P\pi^0}|^2 = 3|A^0|^2 + \tfrac{2}{3}|A^{3/2}|^2 + \tfrac{1}{3}|A^{1/2}|^2. \tag{2.21}$$

Eq. (21) shows that in the *sum* of π^+- and π^0-production off protons no interference between different isospin amplitudes appears. Therefore, even if the correct charges are taken into account in Eq. (20), only such multipoles appear together which according to the Watson theorem should have the same phase.

Using the assumptions of *T*-invariance and one-photon exchange, *Christ* and *Lee* have shown that σ_P must vanish generally, not only in the elastic region and for single pion production, but even if arbitrary final hadron states are produced in a single arm experiment.

From general reasons one expects electromagnetic *T*-violating effects to show up only if a resonance is produced in the final state. No effect is expected in elastic electron scattering [32]. In Eq. (20) only the helicity 1/2 part of the transverse multipole appears, so that the second and third resonances have little chance to contribute to σ_P. If we tentatively assume that the possible *T*-violation invalidates the Watson theorem only for the multipoles M_{1+}, E_{1+} and S_{1+} for $I=3/2$, we can give a simple estimate of the resulting asymmetry A: Neglecting E_{1+} and approximating crudely $\sigma_t+\varepsilon\sigma_l$ by only the $M_{1+}^{3/2}$-contribution, with $\varepsilon\approx 1$ we find

$$A \approx 4\sin(\delta_{S_{1+}}-\delta_{M_{1+}})\,\frac{|\hat{S}_{1+}^{3/2}|}{|M_{1+}^{3/2}|}, \tag{2.22}$$

where $\delta_{S_{1+}}$ and $\delta_{M_{1+}}$ are the phases of $\hat{S}_{1+}^{3/2}$ and $M_{1+}^{3/2}$, respectively. Now we know [2] that $|\hat{S}_{1+}^{3/2}|/|M_{1+}^{3/2}|$ is only of the order of 5...10% and ac-

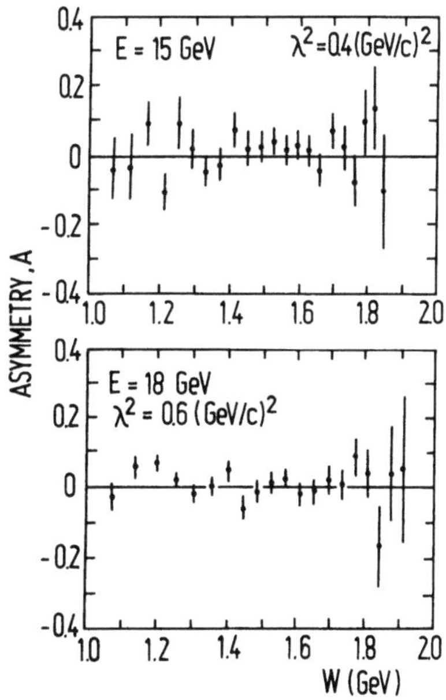

Fig. 10. Example of the SLAC results (*Rock et al.* [33]) for the asymmetry of inelastic electron scattering on a polarized nucleon target

cordingly we expect the asymmetry effect to be small even if the Watson theorem is grossly violated. Results of a recent measurement of A are shown in Fig. 10 (see also Ref. [48]). The data for $\lambda^2 = 0.6\,(\text{GeV}/c)^2$ may indicate a slight asymmetry in the region of the $\Delta(1236)$, but the measurements of the same authors [33] at $\lambda^2 = 1.0$ and 1.4 $(\text{GeV}/c)^2$ do not confirm this result.

It may be that the T-violating interaction obeys a $\Delta I = 0$ rule. In this case the $\Delta(1236)$, being a $I = 3/2$ state, should not show T-violation effects too.

If an asymmetry effect would be found in electroproduction, one had to check whether the asymmetry is due to T-violation or to a breakdown of the one-photon exchange assumption. This can be done by performing an analogous experiment with incident positrons instead of electrons.

III. Theoretical Models for the Virtual Photoproduction Amplitudes

III.1. Introduction

Recent theoretical work on low-energy electroproduction has been performed along four main lines:

1. Predictions of the magnitude and λ^2-dependence of the $(\gamma N N^*)$-form factors from partial-wave dispersion relations,
2. Detailed predictions of the angular distributions from fixed-t-dispersion relations, i.e. calculations of the deviations from a simple Born plus isobar model.
3. Predictions of electroproduction near threshold from current algebra,
4. Quark model predictions.

In the following, we shall concentrate on the first two topics.

The third method will be treated in detail by Dr. *Verzegnassi* [29]. Further details on the first approach will also be given by Dr. *Gutbrod* in his talk at this meeting. For the quark model predictions see e.g. Ref. [34].

III.2. Prediction of the Transition Form Factors

In order to obtain predictions on the magnitude and the λ^2- and W-dependences of the transition form factors, most authors start from the Omnes-Muskhelishvili solution of the partial wave dispersion relation for $\mathcal{M}(W, \lambda^2)$, the multipole amplitude leading to the resonance in question:

$$\mathcal{M}(W, \lambda^2) = \frac{1}{\pi D(W)} \int_L \frac{\mathrm{d}W'}{W'-W} D(W')\, \Delta\mathcal{M}^{\mathrm{L}}(W', \lambda^2). \qquad (3.1)$$

E.g. in the case of the first resonance, $\mathcal{M}(W, \lambda^2)$ stands for one of the multipoles $M_{1+}^{3/2}$, $E_{1+}^{3/2}$ or $S_{1+}^{3/2}$ divided by some kinematical factor which takes care of kinematical singularities. The integral in Eq. (1) is over the left hand singularities in the W-plane, the discontinuities corresponding to these singularities are denoted by $\Delta\mathcal{M}^{\mathrm{L}}(W', \lambda^2)$. $D(W)$ is the usual denominator function, it is taken to be independent of λ^2 and to be determined only by the π-N-scattering phase shift $\alpha(W)$:

$$D(W) = \exp\left\{ -\frac{W-W_0}{\pi} \int_{m+\mu}^{\infty} \mathrm{d}W' \frac{\alpha(W')}{(W'-W_0)(W'-W-i\varepsilon)} \right\}. \qquad (3.2)$$

Thus the λ^2-dependence of $\mathcal{M}(W, \lambda^2)$ is produced by $\Delta\mathcal{M}^{\mathrm{L}}(W', \lambda^2)$ only. The problem is to find a reasonable approximation for $\Delta\mathcal{M}^{\mathrm{L}}(W', \lambda^2)$. Several authors have investigated models for $\Delta\mathcal{M}$.

For the near-by singularities of $\mathcal{M}$, the use of the Born approximation should be reasonable, however, far-away singularities are not expected to be determined by the Born approximation alone. Indeed, if in Eq. (1) one uses the $\Delta\mathcal{M}^{\mathrm{L}}$ calculated from the Born terms for all W', quite bad results are obtained in the case of $M_{1+}^{3/2}$ [34, 35].

Dennery [14] and *Adler* [25] discarded the far-away contributions and kept only the near-by discontinuities in a pole fit to the Born approximation. For low values of λ^2, the multipole $M_{1+}^{3/2}$ obtained in this way is close to the experimental value. $E_{1+}^{3/2}$ and $\hat{S}_{1+}^{3/2}$ come out not so well, but also of the right order of magnitude. For larger values of λ^2, say $\lambda^2 \gtrsim 0.4\,(\mathrm{GeV}/c)^2$, the Adler method gives a $M_{1+}^{3/2}$ which falls off too strongly with λ^2.

Gutbrod and *Simon* [36] used a more sophisticated model for $\Delta\mathcal{M}^{\mathrm{L}}$ in which the far-away contributions are approximated by box graphs. It turns out that such a model quite nicely describes the slower λ^2-dependence of $M_{1+}^{3/2}$ which is observed experimentally, see Fig. 11.

The smallness of the excitation of the $\Delta(1236)$ by the multipoles E_{1+} and S_{1+} can be understood in the context of Eq. (1) from the fact that for $M_{1+}^{3/2}$ the nucleon-exchange contribution is dominant, whereas for $\Delta E_{1+}^{3/2}$ and $\Delta S_{1+}^{3/2}$ the long range pion-exchange term is essential.

Still more refined calculations of the W- and λ^2-dependence of the excitation of the $\Delta(1236)$ have been performed by *Gutbrod* using Bethe-Salpeter techniques [37]. If detailed coincidence experiments make it

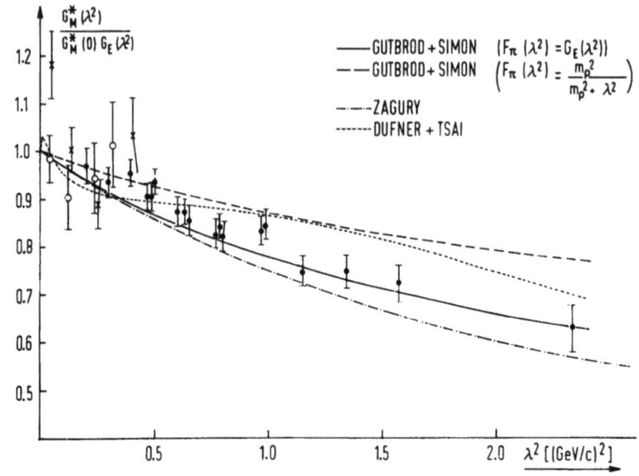

Fig. 11. Predictions for the λ^2-dependence of the magnetic excitation of the $\Delta(1236)$ according to different calculations: *Gutbrod-Simon* [36], *Zagury* [26] and *Adler* [25] (from Ref. [34])

possible to obtain experimental information on $E_{1+}^{3/2}$ and $\hat{S}_{1+}^{3/2}$ it will be interesting to compare the detailed predictions of Ref. [37] for these multipoles to the experimental data.

Within the same framework, the excitation of higher resonances has been calculated by *Walecka* and *Zucker* [38]. In their model for the left hand singularities, $\Delta \mathcal{M}^{\mathrm{L}}$ is calculated from N, π, ω and Δ-exchange diagrams. Definite predictions for the relative magnitude of the excitation of the different resonances by magnetic, electric and scalar multipoles are obtained in this way. So far, no detailed comparison to experimental data has been possible.

III.3. Prediction of Detailed Angular Distributions

If we are interested in making detailed predictions for the angular distribution functions A, B, C and D in Eq. (1.12) for energies in the region of the first resonance, it is important to have reliable predictions for all relevant multipoles, including the nonresonant S-wave multipoles E_{0+} and S_{0+}. For this purpose it is most convenient, instead of trying directly to calculate the left hand discontinuities in the partial wave equations [1], to start with the fixed-t-dispersion relations for the relativistic amplitudes $A_i(s, t, \lambda^2)$ [13, 14]. Calculations for electroproduction along this line have been given by *Zagury* [26], *Adler* [25], *Crawford* [39] and by the author [20, 27].

The partial wave projection of the fixed-t-dispersion relations for the A_i leads to the following coupled set of multipole equations [35, 44, 45, 20]:

$$\begin{aligned}\mathcal{M}_i^{I,J}(W,\lambda^2) = {} & \mathcal{M}_{i\,\mathrm{Born}}^{I,J}(W,\lambda^2) + \frac{P}{\pi}\int_{m+\mu}^{\infty} \frac{\mathrm{d}W'}{W'-W}\,\mathrm{Im}\,\mathcal{M}_i^{I,J}(W',\lambda^2) \\ & + \sum_{I'J'i'} \frac{1}{\pi}\int_{m+\mu}^{\infty} \mathrm{d}W' K_{ii'}^{II'JJ'}(W,W',\lambda^2)\,\mathrm{Im}\,\mathcal{M}_{i'}^{I'J'}(W',\lambda^2).\end{aligned} \tag{3.3}$$

Here $\mathcal{M}_i^{I,J}$ is a multipole amplitude for isospin $I(I=0,\ 1/2$ and $3/2)$ and total angular momentum J. The index i distinguishes the six helicity configurations which are possible for each J (except for $J=1/2$ which has only four possible helicity states). A suitable threshold factor is extracted from the multipoles. $K_{ii'}^{II'JJ'}(W,W',\lambda^2)$ is a complicated kinematical function which can be calculated explicitly once a definite assumption about the possible subtractions in the dispersion relations for the relativistic amplitudes A_i has been made.

With the following provisions, a numerical solution, valid in the low-energy region, can be obtained for the system (3):

1. Only multipoles which have large phases at low energies are kept in the system (3), i.e. the multipoles M_{1+}, E_{1+}, S_{1+}, E_{0+} and S_{0+} for $I=3/2$ and E_{0+} and S_{0+} (eventually also M_{1-}, S_{1-}, M_{2-} and E_{2-}) for $I=1/2$. All other multipoles are not calculated from the system (3) but are taken in the Born approximation.

2. The phases $\delta_{2I,2J}$ of the multipoles $\mathcal{M}_i^{IJ}$, which in the low-energy region are given by the Watson theorem, are assumed to behave at higher energies as follows:

We assume that $\delta_{33}\to\pi$ for high W, and furthermore that δ_1 and δ_3 do not increase too strongly in the inelastic region but rather go back to zero around $W=1.8$ GeV (This permits us to cut off the S-wave integrals around $W=1.8$ GeV). For δ_{13} a Breit-Wigner form is employed. For δ_{11} no definite ideas exist, fortunately its influence seems not to be big.

3. The CDD-ambiguity which arises in the solutions for M_{1+}, E_{1+} and S_{1+} because we assume $\delta_{33}\to\pi$ is resolved by using, if available, experimental values for σ_t and the angular functions D and C at the resonance energy. Observe that for each value of λ^2 only three experimental *points* are needed whereas using these values four *functions* A, B, C and D (which for fixed λ^2 still depend on W and θ) are predicted.

As long as we cannot use experimental data to fix the CDD-ambiguity, one may assume that $M_{1+}^{3/2}$ has no zero and that the zeros of $E_{1+}^{3/2}$, $S_{1+}^{3/2}$ appear at the same value of W where in photoproduction $E_{1+}^{3/2}$ has its zero. As will be seen below in Fig. 12, this prescription works quite well.

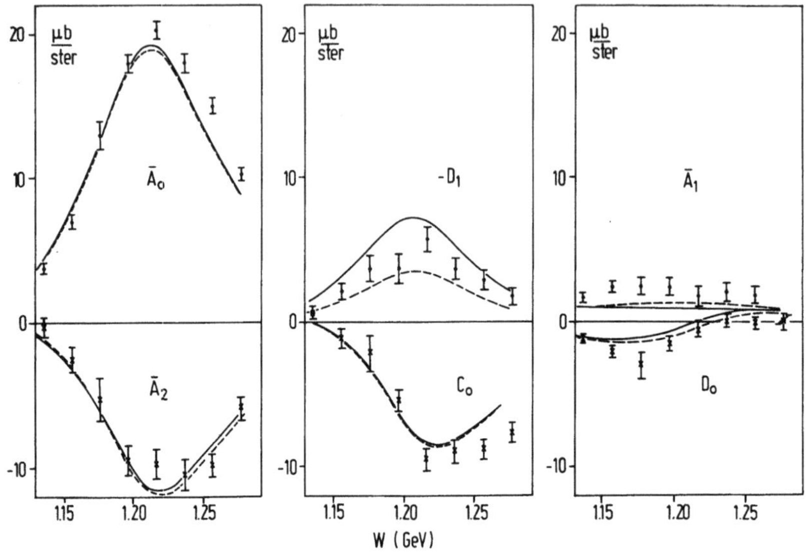

Fig. 12. Experimental results of the DESY-group of *Albrecht et al.* [2] for the angular coefficients for π^0-electroproduction at $\lambda^2 = 0.6\,(\text{GeV}/c)^2$, together with the theoretical predictions of Ref. [20]. The full and dashed curves correspond to the solutions "15" and "15 II" of Ref. [20], respectively (interpolations between the published numbers for 10, 20 and 30 fm^{-2}). The θ-dependence of the A, B etc. of Eq. (1.12) is expanded as follows: $\bar{A} = A + \varepsilon B$ with $\bar{A} = \bar{A}_0 + \bar{A}_1 \cos\theta + \bar{A}_2 \cos^2\theta$, $C = C_0$ and $D = D_0 + D_1 \cos\theta$. Since ε was fixed around $\varepsilon \approx 0.91$, no separation of A and B was possible

4. The kernels in Eq. (3) are calculated using unsubtracted fixed-t-dispersion relations for the A_i except for the subtraction in the amplitude $A_5^{(-)}$ which is demanded by gauge invariance.

No serious convergence difficulties of the multipole expansion are expected as long as $\lambda^2 \lesssim 10\ (\text{GeV}/c)^2$ because of the favorable shape of the Mandelstam spectral regions for $\lambda^2 \neq 0$.

5. The calculation is performed for several plausible values of $F_\pi(\lambda^2)$ and $G_e^N(\lambda^2)$ because these form factors, which enter the calculation through the Born terms, are not yet known experimentally.

Coincidence cross sections calculated with this method are shown in Fig. 12 for $\lambda^2 = 0.6\ (\text{GeV}/c)^2$ together with the recent experimental results obtained by the DESY-Collège de France group [2].

It is of interest to see in what respect this method gives results which go beyond a simple Born + isobar model, i.e. which nonresonant multipoles are predicted to deviate strongly from the Born approximation. Table 1 shows that especially the multipoles E_{0+}, S_{0+}, M_{1-} for $I = 1/2$ and S_{0+} for $I = 3/2$ are found to be quite different from their Born approximation.

Table 1. *Examples of nonresonant multipoles which are predicted to differ strongly from their Born values ($\lambda^2 = 0.6\ (\mathrm{GeV}/c)^2$, $E_\gamma = 315$ MeV)*

Unit: $10^{-2}\mu^{-1}$	$E_{0+}^{1/2}$	$M_{1-}^{1/2}$	$\hat{S}_{0+}^{3/2}$	$\hat{S}_{0+}^{1/2}$
Born	1.2	−.84	−1.4	0.15
Disp. theory	1.8	−.46	−0.6	0.90

The specific predictions of the dispersion calculation can be checked in several ways against experimental data:

1. Near threshold the Born approximation gives

$$\sigma_t^{\pi^0} \ll \sigma_l^{\pi^0} \tag{3.4}$$

whereas the dispersion theory predicts

$$\sigma_t^{\pi^0} \gg \sigma_l^{\pi^0} \tag{3.5}$$

because $\hat{S}_{0+}^{\pi^0}$ is strongly reduced by the dispersion contributions.

2. Information on the small multipoles is contained in the angular coefficients A_1 and D_0. (We write $A = A_0 + A_1 \cos\theta + A_2 \cos^2\theta + \cdots$ and $D = D_0 + D_1 \cos\theta + \cdots$ etc.), e.g.

$$D_0^{\pi^0} \sim \mathrm{Re}\{E_{0+}\hat{S}_{1-}^* + (M_{1-} - M_{1+} + 3E_{1+})\hat{S}_{0+}^* + \cdots\}. \tag{3.6}$$

3. As is well-known from photoproduction, the ratio C_0/A_2 *off* resonance gives information on M_{1-} (*at* resonance it is sensitive to E_{1+}).

4. The coefficient D_1 of the transverse-longitudinal interference term is very sensitive to the multipole $\hat{S}_{1+}$ through the product $\mathrm{Re}(M_{1+}\hat{S}_{1+}^*)$.

5. In the special case $\lambda^2 = 0$ the calculation should give good results for photoproduction. Since for photoproduction multipole analyses of the experimental data are available [40–42], a detailed check of the multipole predictions is possible for $\lambda^2 = 0$. It turns out that the deviations from the Born approximation for the multipoles $E_{0+}^{1/2}$ and $M_{1-}^{1/2}$ which have been established by the multipole analyses are correctly predicted from Eq. (3). However, the deviation of E_{0+}^0 and $M_{1-}^{3/2}$ is not found in a straightforward way from the dispersion calculation of Eq. (3). It is not clear which of the assumptions made in solving the system (3) is to be blamed for this.

For increasing values of λ^2 the solutions of Eq. (3) show the following tendencies: For energies in the resonance region, all multipoles fall quite strongly with λ^2, except for $M_{1+}^{3/2}$ and $E_{0+}^{1/2}$ which fall somewhat slower than ϱ-dominance, and except for $M_{1-}^{1/2}$ and $\hat{S}_{1-}^{1/2}$ which remain almost λ^2-independent. So $M_{1-}^{1/2}$ and $\hat{S}_{1-}^{1/2}$ become important for $\lambda^2 \gtrsim 2\ (\mathrm{GeV}/c)^2$.

We finally remark that the relation $L_{0+} \approx E_{0+}$ which can be proved at the unphysical point $|\boldsymbol{k}| = 0$ [21] turns out to be completely wrong in the physical region. This is possible because the point $|\boldsymbol{k}| = 0$ moves rapidly away from the physical region as λ^2 increases. E.g. for $\lambda^2 = 1\,(\text{GeV}/c)^2$ even at threshold we have already $|\boldsymbol{k}| \approx 8\,\mu$.

IV. Conclusions

We saw that electroproduction experiments in the resonance region can, in principle, give information on $(\gamma N N^*)$-transition form factors, the pion form factor F_π, the neutron electric form factor G_e^N and on the axial nucleon form factor G_A. At present, electroproduction is the only practical source of information on F_π in the spacelike region and on the transition form factors. However, several difficulties are involved:

Most easy is the determination of the pion form factor and of the dominant $(\gamma N \Delta)$-form factor C_3. The form factors for the non-dominant transitions to the $\Delta(1236)$, C_4 and C_5, are already difficult to define because these transitions have a quite unsymmetrical shape of their energy dependence. For the determination of G_e^N electroproduction does not seem to be much superior to electron-deuteron scattering because unknown dispersion effects make the extraction of G_e^N from electroproduction data ambiguous.

Of fundamental importance is the clear-cut time-reversal test in electroproduction which was described in Sec. II.7.

The theoretical models which aim to predict the λ^2-dependence of the $(\gamma N \Delta)$-transition form factors showed that it is crucial to include assumptions on the short-range forces in order to get the observed λ^2-dependence of $M_{1+}^{3/2}$.

The detailed prediction of the angular distributions from fixed-t-dispersion relations agrees reasonably well with the data. When more experimental data from coincidence experiments (possibly for different values of ε) become available it will be interesting to see whether the λ^2-dependences of the different multipoles are predicted correctly.

The author is grateful to Drs. *F. Brasse*, *J. Drees*, *F. Gutbrod*, *D. Schwela*, *C. Verzegnassi* and *H. Wessel* for interesting discussions. Thanks are also due to *H. Wessel* for calculating the angular distributions shown in Fig. 12.

References

1. For a recent review, see: *Clegg, A.B.:* Proc. of the 4th Intern. Symposium on Electron and Photon Interactions at High Energies, Daresbury 1969.
2. *Albrecht, W., et al.:* Nucl. Phys. B **25**, 1 (1971).
3. *Nuthakki, V.P.R., et al.:* Daresbury preprint DNPL/P 50 (Sept. 1970).
4. *Amaldi, E., et al.:* Nuovo Cimento **65** A, 377 (1970).

5. *Kajiura, N.*, *et al.*: Lettere Nuovo Cim. **3**, 714 (1970).
6. *Moritz, J.*: Thesis Karlsruhe 1970.
7. *Hand, L. N.*: Phys. Rev. **129**, 1834 (1963).
8. *Gilman, F. J.*: Phys. Rev. **167**, 1365 (1968).
9. *Gourdin, M.*: Nuovo Cimento **21**, 1094 (1961).
10. *Akerlof, C. W.*, *et al.*: Phys. Rev. **163**, 1482 (1967).
11. *Christ, N.*, *Lee, T. D.*: Phys. Rev. **143**, 1310 (1966).
12. *v. Gehlen, G.*: Phys. Rev. **118**, 1455 (1960), Nucl. Phys. B **26**, 141 (1971).
13. *Fubini, S.*, *Nambu, Y.*, *Wataghin, V.*: Phys. Rev. **111**, 329 (1958).
14. *Dennery, Ph.*: Phys. Rev. **124**, 2000 (1961).
15. *Zagury, N.*, *Teixeira, A. F. F.*: Nuovo Cimento **61** A, 83 (1969).
16. *Bartel, W.*, *et al.*: Phys. Letters **27** B, 660 (1968); **28** B, 148 (1968).
17. *Gourdin, M.*, *Salin, Ph.*: Nuovo Cimento **39**, 591 (1965).
18. *Dufner, A. J.*, *Tsai, Y. S.*: Phys. Rev. **168**, 1801 (1968).
19. *Loubaton, J. P.*: Nuovo Cimento **39**, 591 (1965).
20. *v. Gehlen, G.*: Nucl. Phys. B **9**, 17 (1969); B **20**, 102 (1970).
21. *Bjorken, J. D.*, *Walecka, J. D.*: Ann. Phys. **38**, 35 (1966).
22. *Bloom, E. D.*, *et al.*: preprint SLAC-PUB-796 (Sept. 1970).
23. *Block, M. M.*, *et al.*: Phys. Rev. **169**, 1074 (1968).
24. *Mistretta, C.*, *et al.*: Phys. Rev. Letters **20**, 1523 (1968).
25. *Adler, S. L.*: Ann. Phys. **50**, 189 (1968).
26. *Zagury, N.*: Phys. Rev. **145**, 1112 (1966), Nuovo Cimento **52** A, 506 (1967).
27. *v. Gehlen, G.*: preprint Bonn Univ. PI 2–80 (Aug. 1970).
28. *De Tollis, B.*, *Nicolò, F.*: Nuovo Cimento **48**, 281 (1967).
29. *Verzegnassi, C.*: This volume.
30. *Adler, S. L.*, *Gilman, F.*: Phys. Rev. **152**, 1460 (1960).
31. *v. Gehlen, G.*, *Schmidt, M. G.*: Nucl. Phys. B **20**, 173 (1970).
32. *Bernstein, J.*, *Feinberg, G.*, *Lee, T. D.*: Phys. Rev. **139**, B 1650 (1965).
33. *Rock, S.*, *et al.*: Phys. Rev. Letters **24**, 748 (1970).
34. *Gutbrod, F.*: Proc. of the 4th Intern. Symposium on Electron and Photon Interactions at High Energies, Daresbury 1969.
35. *Schwela, D.*, *Rollnik, H.*, *Weizel, R.*, *Korth, W.*: Z. Physik **202**, 452 (1967).
36. *Gutbrod, F.*, *Simon, D.*: Nuovo Cimento **51** A, 602 (1967).
37. — Preprints DESY 69/22, DESY 69/33 (1969).
38. *Walecka, J. D.*, *Zucker, P. A.*: Phys. Rev. **167**, 1479 (1968).
39. *Crawford, R. L.*: Nucl. Phys. B **28**, 573 (1971).
40. *Nölle, P.*, *Pfeil, W.*, *Schwela, D.*: Nucl. Phys. B **26**, 461 (1971).
41. *Nölle, P.*, *Pfeil, W.*: Nucl. Phys. B **31**, 1 (1971).
42. *Berends, F. A.*, *Weaver, D. L.*: Nucl. Phys. B **30**, 575 (1971).
43. *Breidenbach, M.*: MIT Report No. 2098–635.
44. *Schmidt, W.*, *Schwiderski, G.*: Fortschr. Phys. **15**, 393 (1967).
45. *Donnachie, A.*, *Shaw, G.*: Ann. Phys. (N.Y.) **37**, 333 (1966).
46. *Titov, Y. I.*, *et al.*: Soviet J. Nucl. Phys. **11**, 525 (1970), JETP Letters **12**, 129 (1970).
47. *Omelaenko, A. S.*: Soviet J. Nucl. Phys. **11**, 237 (1970).
48. *Chen, J. R.*, *et al.*: Phys. Rev. Letters **21**, 1279 (1968).
49. *Bartel, W.*, *et al.*: Phys. Letters **35 B**, 181 (1971).
50. *Driver, C.*, *et al.*: Phys. Letters **35 B**, 77 (1971).
51. *Brown, C. N.*, *et al.*: Phys. Rev. Letters **26**, 987 (1971).
52. *Kummer, P. S.*, *et al.*: Daresbury preprint DNPL/P 67 (1971).

Dr. *G. von Gehlen*
Physikalisches Institut der Universität
5300 Bonn, Germany

Experimental Data on Photoproduction of Pseudoscalar Mesons at Intermediate Energies

H. FISCHER

Contents

1. Introduction

Photoproduction of pseudoscalar mesons from nucleons in the resonance region has been extensively studied during the last two decades. The investigations have mainly been concentrated on the following processes:

$$\begin{aligned}
&\gamma + p \to \pi^+ + n \\
&\gamma + p \to \pi^0 + p \\
&\gamma + n \to \pi^- + p \\
&\gamma + p \to \eta + p \\
&\gamma + p \to K^+ + \Lambda \\
&\gamma + p \to K^+ + \Sigma^0
\end{aligned}$$

The quantities which have been determined from experiment are:

a) The *Differential Cross Section* $\mathrm{d}\sigma/\mathrm{d}\Omega$ for photoproduction of mesons from unpolarized protons using unpolarized photons. Following the notation of *R. L. Walker* [68] this cross section can be written in terms of helicity amplitudes H_i:

$$\frac{\mathrm{d}\sigma}{\mathrm{d}\Omega}(\theta) = \frac{1}{2}\,\frac{q}{k}\,(H_1^2 + H_2^2 + H_3^2 + H_4^2)$$

k and q are the momenta of photon and pion in the CM system.

b) The *Polarization* $P(\theta)$ of the recoil nucleon when the initial particles are unpolarized

$$P(\theta)=\frac{(\mathrm{d}\sigma/\mathrm{d}\Omega)\uparrow-(\mathrm{d}\sigma/\mathrm{d}\Omega)\downarrow}{(\mathrm{d}\sigma/\mathrm{d}\Omega)\uparrow+(\mathrm{d}\sigma/\mathrm{d}\Omega)\downarrow}=\frac{q}{k}\,\frac{1}{\mathrm{d}\sigma/\mathrm{d}\Omega}\,\mathrm{Im}(H_1H_3^*+H_2H_4^*)$$

$(\mathrm{d}\sigma/\mathrm{d}\Omega)\uparrow$ and $(\mathrm{d}\sigma/\mathrm{d}\Omega)\downarrow$ are cross sections for meson production with spin of the recoil nucleon oriented up and down perpendicular to the production plane.

Positive values of polarization are defined in the direction $(\boldsymbol{k}\times\boldsymbol{q})$.

c) The *Asymmetry Ratio* $\Sigma(\theta)$ for photoproduction from unpolarized protons using linearly polarized photons

$$\Sigma(\theta)=\frac{(\mathrm{d}\sigma/\mathrm{d}\Omega)_\perp-(\mathrm{d}\sigma/\mathrm{d}\Omega)_\parallel}{(\mathrm{d}\sigma/\mathrm{d}\Omega)_\perp+(\mathrm{d}\sigma/\mathrm{d}\Omega)_\parallel}=\frac{q}{k}\,\frac{1}{(\mathrm{d}\sigma/\mathrm{d}\Omega)}\,\mathrm{Re}(H_1H_4^*-H_2H_3^*)$$

where $(\mathrm{d}\sigma/\mathrm{d}\Omega)_\perp$ and $(\mathrm{d}\sigma/\mathrm{d}\Omega)_\parallel$ are the cross sections for photon polarization parallel and perpendicular to the production plane.

Another quantity for which there are up to now no data published is

d) The *Asymmetry Ratio* $T(\theta)$ for photoproduction from polarized protons by unpolarized photons

$$T(\theta)=\frac{\uparrow(\mathrm{d}\sigma/\mathrm{d}\Omega)-\downarrow(\mathrm{d}\sigma/\mathrm{d}\Omega)}{\uparrow(\mathrm{d}\sigma/\mathrm{d}\Omega)+\downarrow(\mathrm{d}\sigma/\mathrm{d}\Omega)}=\frac{q}{k}\,\frac{1}{\mathrm{d}\sigma/\mathrm{d}\Omega}\,\mathrm{Im}(H_1H_2^*+H_3H_4^*)$$

where $\uparrow(\mathrm{d}\sigma/\mathrm{d}\Omega)$ and $\downarrow(\mathrm{d}\sigma/\mathrm{d}\Omega)$ are the cross sections for meson production from protons polarized up and down in the direction perpendicular to the production plane.

Polarized targets are being built in different laboratories, so that there will be measurements of $T(\theta)$ in the near future. A sizeable negative asymmetry for $\gamma+p\to\pi^++n$ has already been observed in a first test run at Bonn [90] for a photon energy of 1.125 GeV at a CM angle of 45°.

In the following we will limit the discussion to intermediate photon energies from about 0.5 to 2.5 GeV. The activities in this field during the last year can best be shown by the list of papers which have been presented at the Kiev Conference on High Energy Physics in September 1970 (Table 1).

We will not discuss these experiments in detail. I rather will try to give a survey over the existing experimental data. Before going to the experimental part I will briefly show the different approaches which have been made to analyse the experimental results in the resonance region.

Table 1

Experiment			Angular Range	Energy Range (GeV)
$\gamma+p\to\pi^++n$	$d\sigma/d\Omega$	Tokyo [7]	180°	0.3 –1.2
	$d\sigma/d\Omega$	Cornell [6]	180°	0.6 –7.0
	$d\sigma/d\Omega$	Bonn [8]	180°	0.6 –1.3
$\gamma+p\to\pi^0+p$	$P(\theta)$	Bonn [24]	45°–148°	0.5 –1.2
	$\Sigma(\theta)$	Kharkov [27]	90°	0.25–0.6
	$\Sigma(\theta)$	Frascati [26]	90°	0.28–0.6
$\gamma+p\to\eta+p$	$P(\theta)$	Caltech [42]	90°	0.8 –1.1
$\gamma+p\to K^++\Lambda$	$d\sigma/d\Omega$	Orsay [55]	67°–135°	0.9 –1.7
	$d\sigma/d\Omega$	Bonn [51]	6°–90°	1.3 –1.34
$\gamma+p\to K^++\Sigma^0$	$d\sigma/d\Omega$	Orsay [55]	75°–138°	1.1 –1.8
	$d\sigma/d\Omega$	Bonn [51]	12°–84°	1.41–1.45

2. Analysis of Meson Photoproduction Data in the Intermediate Energy Region

Photoproduction of pions at photon energies below 0.5 GeV is dominated by the so called "first" resonance which has its maximum around 0.3 GeV. At higher energies the "second" and "third" resonance show up as flat bumps on a comparatively large nonresonant background at photon energies of 0.7 and 1.0 GeV. In Fig. 1 the total cross sections for $\gamma+p\to\pi^++n$, $\gamma+p\to\pi^0+p$ and $\gamma+n\to\pi^-+p$ are plotted versus the lab photon energy. The curves give the results of a series of fits to the experimental data [28, 67, 68, 70].

The first resonance is due to the excitation of one single resonance namely the $P_{33}(1236)$. At higher energies the situation is more complex, because several resonances can contribute to the photoproduction amplitude. Fig. 2 gives a summary of the pion nucleon resonances known from pion nucleon scattering according to the most recent edition of "Particles Properties" of the Particle Data Group [89]. The corresponding lab photon energies have been calculated using the total CM energies written in brackets for each resonance.

It is well known that the second and third resonance are mainly due to excitation of the $D_{13}(1520)$ and the $F_{15}(1688)$. The contributions of the other resonances which exist in the same energy regions are at least small. Whether some of them are excited at all in photoproduction is still under discussion. Little is known about the resonance behaviour at energies above 1.2 GeV. There are not enough data, as we will see later, to carry out a decent analysis.

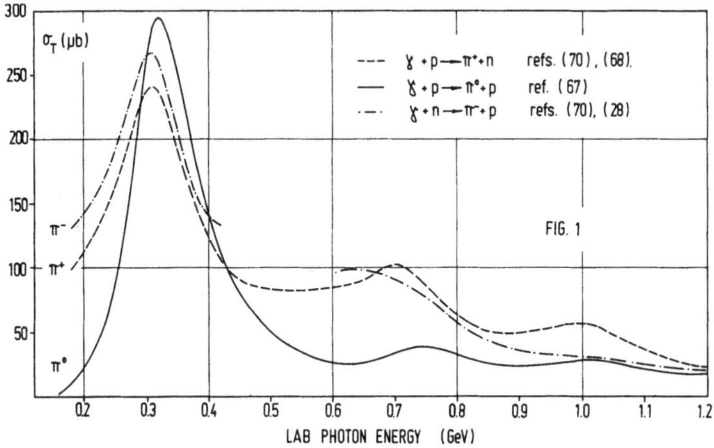

Fig. 1. Total cross sections for photoproduction of pions from nucleons. Data are from Refs. [28, 67, 68, 70]

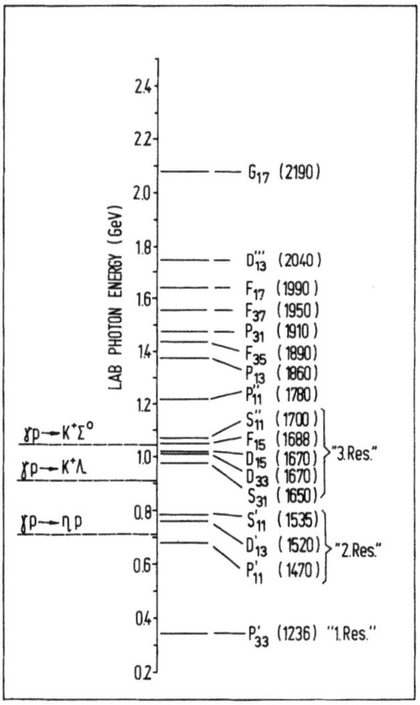

Fig. 2. Summary of the known pion nucleon resonances [87] and appropriate lab photon energies. The broken lines indicate the lab threshold energies for η and K^+ production

The lab photon energies at threshold for η and K^+ production are also marked in the energy level scheme of Fig. 2, to give an idea which resonances can contribute to these processes. Which of them in fact decay into ηp, $K^+\Lambda$ and $K^+\Sigma^0$ will be discussed later together with the experimental data.

	$\gamma+p\to\eta+p$	$\gamma+p\to K^++\Lambda$	$\gamma+p\to K^++\Sigma^0$
Lab photon energy at threshold (MeV)	710	910	1046

There are two general approaches to the calculation of the photoproduction amplitude in the resonance region.

The first approach is based on fixed t dispersion relations and the Watson theorem, according to which the phase of a multipole amplitude in photoproduction is equal to the pion nucleon scattering phase shift if the energy is below inelastic threshold ($E_\gamma = 332$ MeV). For this latter reason calculations of this type have been limited to the region of the first resonance. The different approximations and solutions used have been discussed in detail by *G. Höhler* [87] and *H. Rollnik* [88].

The second approach is the phenomenological analysis. The aim in such an analysis is to determine the photoproduction amplitudes from measured quantities using only general theoretical properties (e.g. unitarity) but no specific dynamical model.

A multipole analysis of this kind without further assumptions could so far only be achieved in the first resonance region [61, 86].

At higher energies an attempt for an energy independent multipole analysis has been reported by *F. A. Berends* and *A. Donnachie* [62, 63]. In a first approximation they calculated the multipoles by dispersion relations using the Watson theorem also above inelastic threshold. The obtained amplitudes were then used as starting point for the analysis. Final values for the multipoles were eventually gained by fit to the experimental data. In order to get unique solutions the nonresonant multipoles were kept unchanged whereas the resonant amplitudes were allowed to vary within reasonable limits.

All other phenomenological approaches use a specific parametrisation of those multipoles which are assumed to be important. In all cases the multipole amplitude $M(W)$ for a specific spin-parity-isospin channel is decomposed into an s-channel resonance part $M_{\text{res}}(W)$ and a background amplitude $M_{\text{BG}}(W)$.

$$M(W) = M_{\text{res}}(W) + M_{\text{BG}}(W).$$

W is the total energy in the C.M.system.

The resonance parts of the different approaches differ only slightly, whereas the assumptions for the background amplitude are quite different.

The main features of the various calculations which have been carried out can be classified using the following schemes.

Parametrisation for $M_{res}(W)$

A	$M_{res}(W) = (\text{B.W.})$
B	$M_{res}(W) = e^{i\varphi} \times (\text{B.W.})$
C	$M_{res}(W) = \text{const} \times f(W)$

where (B.W.) stands for a Breit Wigner amplitude. φ is a constant phase relative to the nonresonant background and $f(W)$ is the elastic π-N-scattering partial wave amplitude. A is a pure resonance formula while in B and C parts of the background are already included.

Assumptions for $M_{BG}(W)$

1	$M_{BG}(W)$ = smooth function treated point by point
2	$M_{BG}(W)$ = smooth function using assumed energy dependence
3	$M_{BG}(W)$ = elementary t- and u-channel exchange (Isobar Model)
4	$M_{BG}(W)$ = reggeized t- (or u-) channel exchange (Generalized Interference Model)

Table 2–4 indicate the amplitude assumptions and the angle and energy range of the most recent analysis, together with the authors of the different calculations.

Table 2

$\gamma + N \to \pi + N$			
Authors	$M_{res}(W)$	$M_{BG}(W)$	Energy Range (GeV)
T. Yamaki (1967) Ref. [65]	C	1 + C.L.[a]	≦0.9
Y. C. Chau, R. G. Dombey and *R. G. Moorhouse* (1967) Ref. [66]	B	2	0.535–0.85
W. Schmidt and *G. Schwiderski* (1968) Ref. [64]	C	1 + I.A.[b]	0.6–1.2
W. Pfeil, Kim Yong Taik (1968) Ref. [68]	A	3	≦1.2
R. L. Walker (1969) Ref. [69]	A	1	≦1.2
R. G. Moorhouse and *W. A. Rankin* (1970) Ref. [69]	B	2	0.6–0.825
A. Proia and *F. Sebastiani* (1970) Ref. [71]	B	2	0.54–0.82

[a] Chew-Low-Amplitudes.
[b] P_{33}-Isobar-Approximation.

Table 3

$\gamma+N\to\eta+N$			
Authors	$M_{res}(W)$	$M_{BG}(W)$	Energy Range (GeV)
S. R. Deans and *W. G. Holladay* (1967) Ref. [72]	A	3	$\leqq 1.1$
S. K. Srinivasan, P. Achuthan and *R. N. Sarkar* (1969) Ref. [73]	A	3	0.7–1.3
H. G. Beisenherz (1969) Ref. [74]	A	3	0.7–1.4
R. P. Bajpai and *A. Donnachie* 1969) Ref. [75]	A	4	$\leqq 2.0$

Table 4

$\gamma+N\to K+\gamma$			
Authors	$M_{res}(W)$	$M_{BG}(W)$	Energy Range (GeV)
H. Thom (1966) Ref. [49]	A	3	$\leqq 1.4$
S. Orito (1969) Ref. [76]	B	2	$\leqq 1.4$
W. Schorsch, J. Tietge and *W. Weilnböck* (1979) Ref. [77]	B	2	$\leqq 2.0$
V. Meyer zu Hörste and *W. Pfeil* (1970)	—	4 only	1.2–16.0
Ref. [78]	A	3	$\leqq 1.7$
F. M. Renard and *Y. Renard* (1970) Ref. [79]	A	3	$\leqq 1.8$

3. Single Pion Photoproduction

Most of the work on pseudoscalar mesons has been done on pion photoproduction. The experimental results for photon energies below 1.5 GeV which have been obtained up to 1966 are listed in the data compilation of *J. T. Beale, S. D. Eklund* and *R. L. Walker* [1]. A compilation of photoproduction data above 1.2 GeV has just been published by *P. Joos* [91].

3.1. Positive Pion Production in $\gamma+p\to\pi^{+}+n$

A summary of π^{+}-production from hydrogen is shown in Fig. 3. Angular distributions of the differential cross section have been measured mainly at Caltech [1] for photon energies between 0.5 and 1.27 GeV at an average distance of about 20 MeV. The angular range is in general from 6° to 165° except for energies below 0.6 GeV where the measurements reach only down to 45°. New data at forward angles have been measured at Orsay [2] for energies between 0.3 and 0.75 GeV and at angles from 0° to 130°. This new set of measurements shows good agreement with the existing data [1] and also with the recent Bonn measurements [83] around the first resonance.

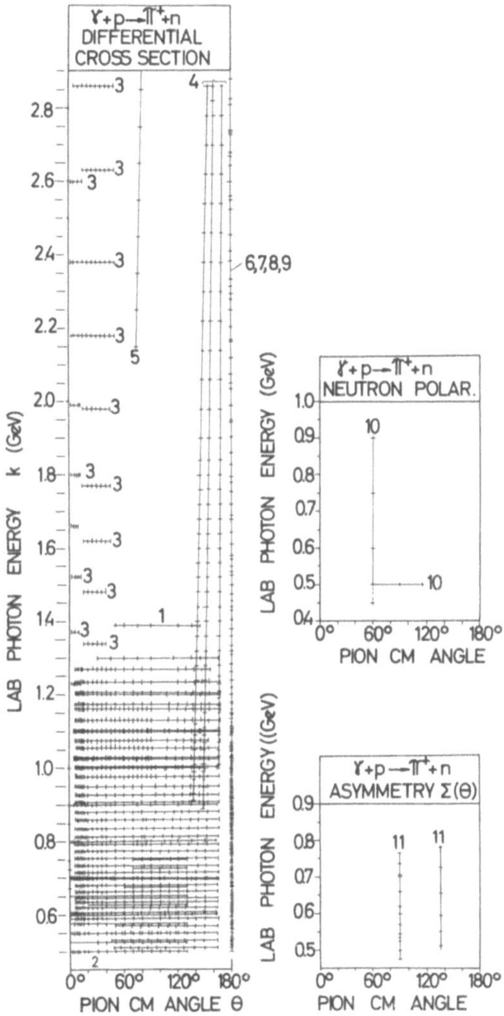

Fig. 3. Summary of experimental data on $\gamma + p \to \pi^+ + n$. The energies and angles at which measurements of differential cross section, recoil nucleon polarization and asymmetry have been made are denoted by a cross in the θ-k-plane. Angular distributions are shown by horizontal lines, excitation curves by vertical lines. The numbers give the different experiments according to the list of references. The angular distributions between 0.5 and 1.3 GeV are from Refs. [1, 2]

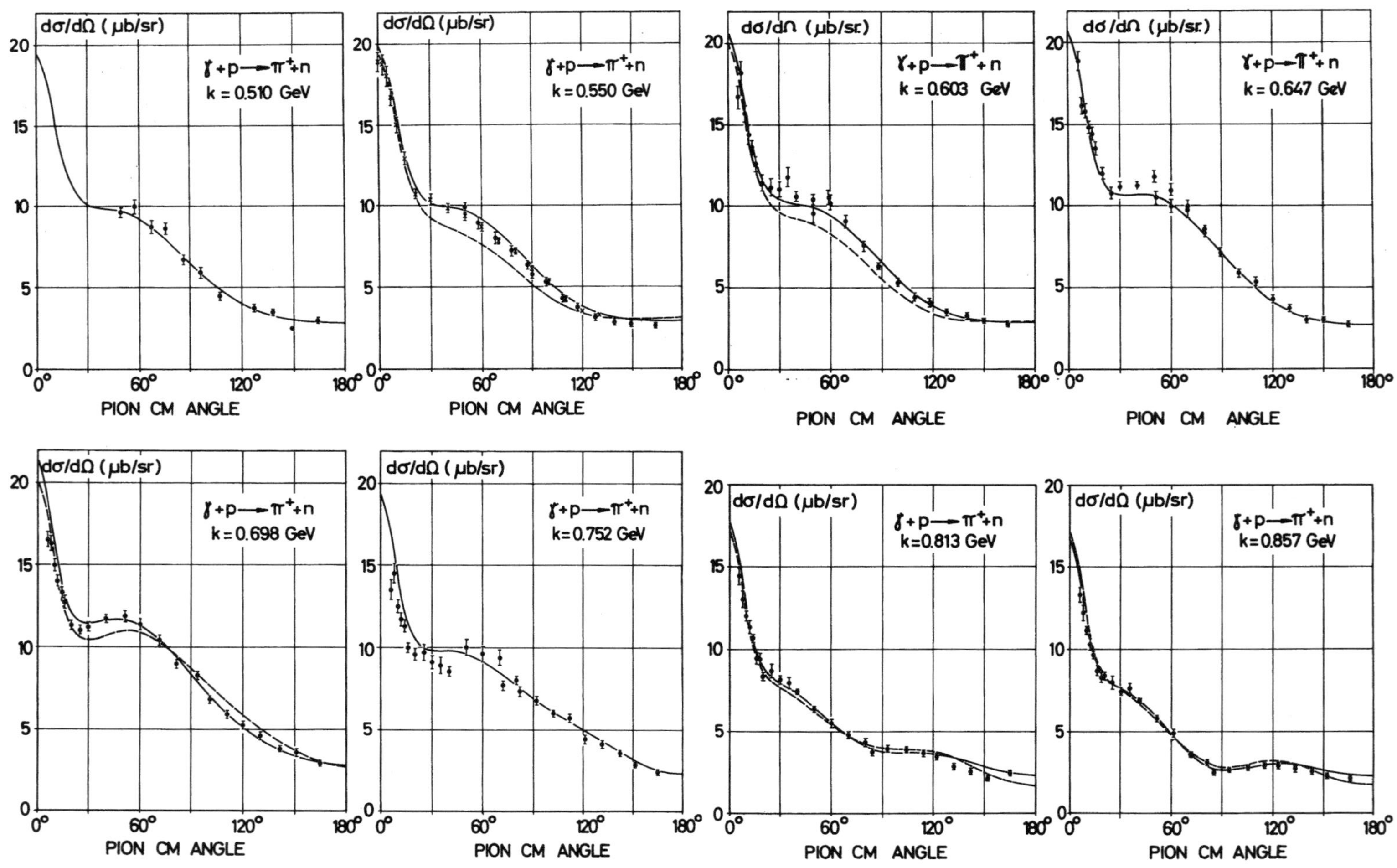
dσ/dΩ (μb/sr)
γ+p → π⁺+n
k = 0.510 GeV
k = 0.550 GeV
k = 0.603 GeV
k = 0.647 GeV
k = 0.698 GeV
k = 0.752 GeV
k = 0.813 GeV
k = 0.857 GeV
PION CM ANGLE

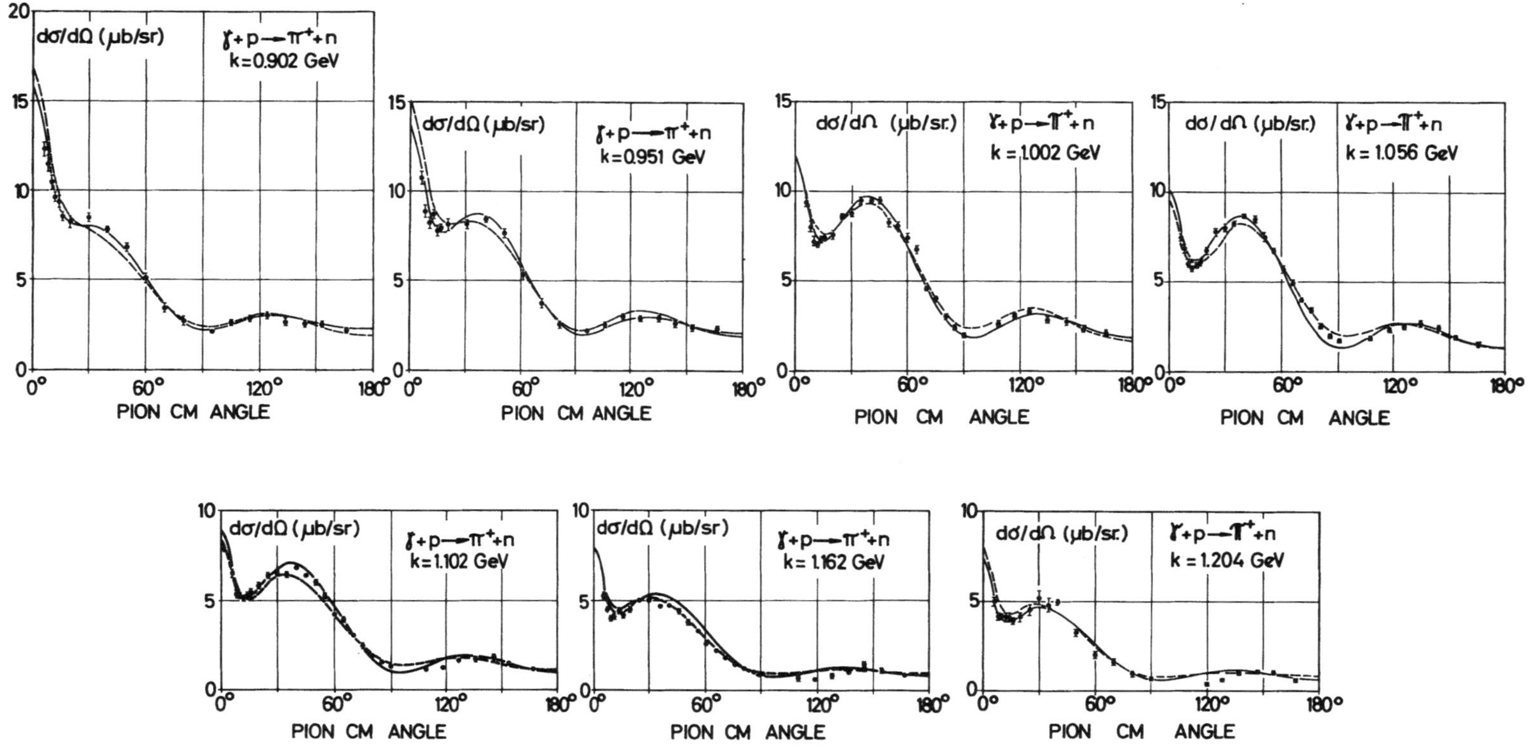

Fig. 4. Angular distributions for $\gamma + p \to \pi^+ + n$. The data are from Ref. [1] and Ref. [2]. The solid lines are fit curves from *R. L. Walker* [68] the broken lines from *Kim Yong Taik* and *W. Pfeil* [67]

A sample of 15 angular distributions is shown in Fig. 4. Only a slight enhancement around 0.7 GeV indicates the contribution of the "second resonance" $D_{13}(1520)$, whereas the contribution of the "third resonance" $F_{15}(1688)$ shows up more strongly around 1 GeV. The solid lines are fit curves from *R. L. Walker* [67] and the broken lines from the analysis of *Kim Yong Taik* [68].

There exist only few data on backward production above 165°. Cross sections at 180° have been measured at Stanford [81] at energies between 0.6 and 0.84 GeV. According to *H. A. Thiessen* [82] there is some evidence that the normalisation of this data might be incorrect. For the purpose of a detailed multipole analysis it is necessary to have more and improved data in the backward region. On the other hand measurements at 0° and 180° are of special interest, because the photoproduction amplitude in these directions takes a simple form, e.g. only one helicity amplitude contributes to the differential cross section.

Therefore new measurements of the 180° cross section have been carried out recently at four different laboratories. Data are reported from Tokyo [7] for a photon energy range of 0.3–1.2 GeV, from Bonn [8] for 0.6–1.3 GeV, from Orsay-Daresbury [9] for 0.45–2.9 GeV and from Cornell [6] at 0.65–7.1 GeV.

The results up to 1.3 GeV are plotted in Fig. 5a. There are considerable deviations in absolute value between the different experiments, but they all show a shoulder at the position of the second resonance and a flat bump in the region of the third resonance. The same energy behaviour is predicted by the *Walker* [67] analysis (full line) and the calculations of *Kim Y. Taik* and *W. Pfeil* [68] (broken line).

At low energies there is good agreement between the Tokyo and Orsay results on one hand and the Cornell and Bonn results on the other hand which are systematically lower. Above 1 GeV there is an increasing agreement between Orsay and Bonn, whereas the data from Tokyo and Cornell agree at considerable higher values. This large discrepancy is probably due to deviations in momentum calibration. Backward produced pions have low momenta and therefore suffer high energy losses in matter. Due to the adverse kinematics in backward direction a small momentum change gives rise to a large shift in photon energy. For all data are more or less preliminary, more detailed corrections to the data may soon bring a better agreement.

Polarization $P(\theta)$ of the recoil neutron has been investigated at Bonn [10]. Fig. 5b shows an excitation curve at 60° between 0.5 and 0.9 GeV photon energy and also an angular distribution at 0.5 GeV. The data show good agreement with the predictions of Refs. [67, 68]. It is worthwhile to remind that these data have not been included in both of the analyses.

Measurements of the asymmetry parameter $\Sigma(\theta)$ have been published from Stanford [11]. Fig. 5c shows the energy dependence at 90° and at 135°. The curves again are fits to the data from Refs. [67, 68].

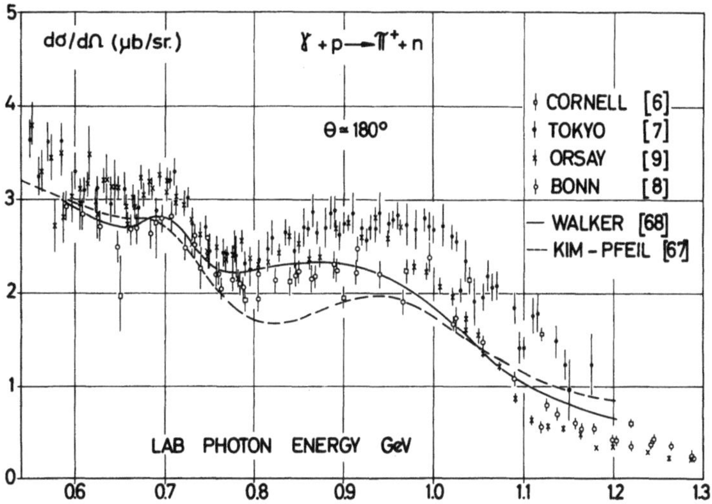

Fig. 5a. Energy dependence of 180° cross section for $\gamma + p \to \pi^+ + n$ showing the data from Cornell [6], Tokyo [7], Bonn [8] and Orsay [9]. The fit curves are from Refs. [67, 68]

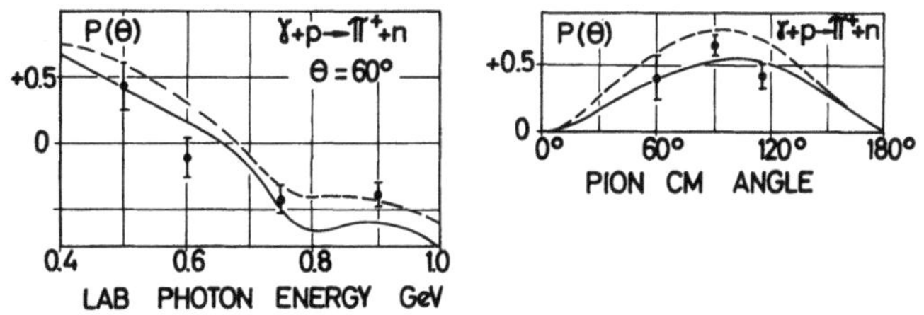

Fig. 5b. Polarization of the recoil neutron for $\gamma + p \to \pi^+ + n$. Data are from Ref. [10]. Solid curve from Ref. [68], broken line from Ref. [67]

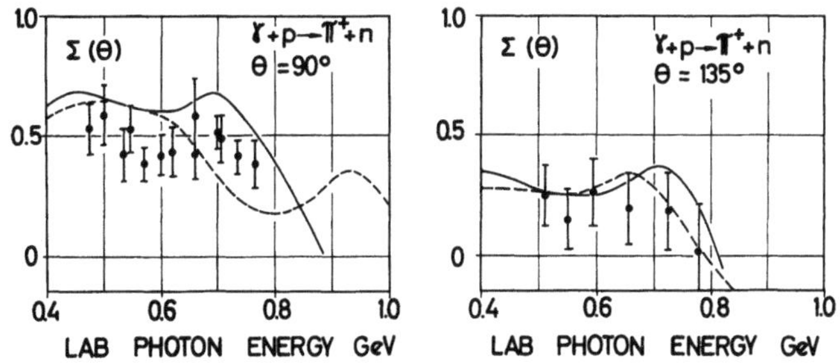

Fig. 5c. Asymmetry parameter for linearly polarized photons at 90° and 135°. Data are from Ref. [11]. The curves are from Refs. [68] (solid line) and [67] (broken line)

In the energy region above 1.4 GeV only forward and backward data have been reported except the excitation curve from CEA [5] at central angles above 2.1 GeV.

Angular distributions at forward angles between 2.5° and 50° have been measured at DESY [63] up to 2.9 GeV. The data are plotted in Fig. 6 versus photon energy to enable a comparison with the three

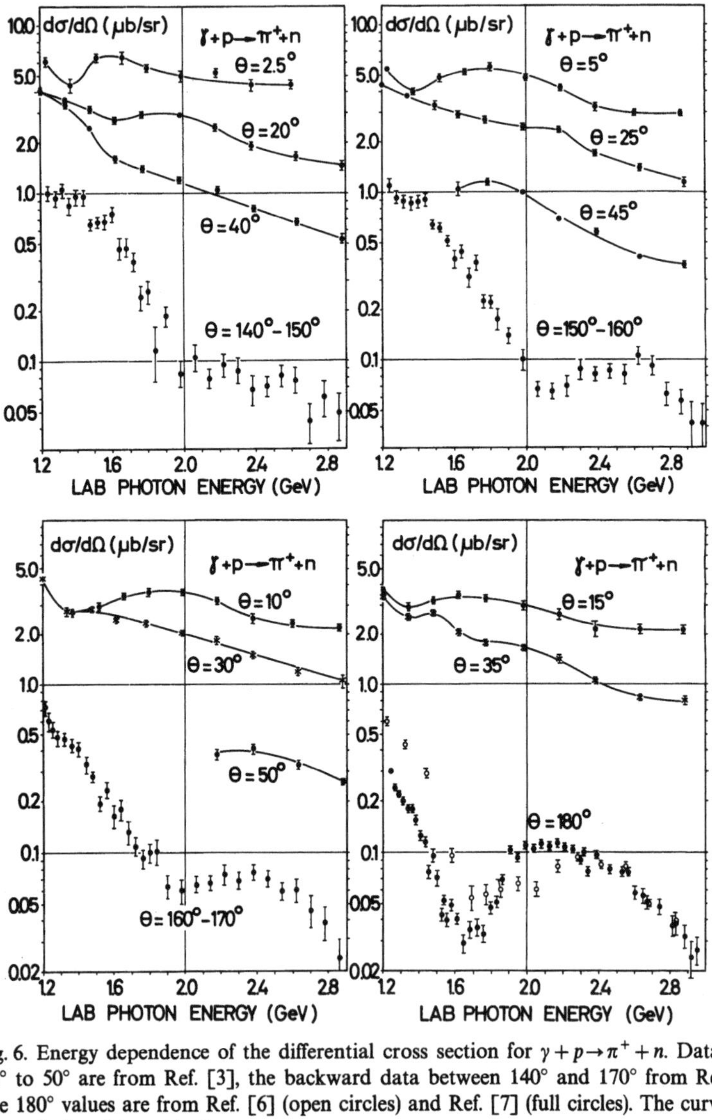

Fig. 6. Energy dependence of the differential cross section for $\gamma+p\rightarrow\pi^{+}+n$. Data from 2.5° to 50° are from Ref. [3], the backward data between 140° and 170° from Ref. [4]. The 180° values are from Ref. [6] (open circles) and Ref. [7] (full circles). The curves are drawn by hand to guide the eye

excitation curves from CEA [4] taken between 140° and 170° and also with the already mentioned 180° data of Orsay and Cornell. The lines along the DESY data are drawn by hand to guide the eye. The main features in this energy region are the following: There is a steep forward peak due to the pion pole term contribution which shows a flat decrease with energy. At backward angles we have a steep decrease of the cross section which is superimposed by resonance contributions. Possible candidates are the $F_{37}(1920)$ around 1.5 GeV and the $\Delta(2420)$ around 2.4 GeV.

3.2. Neutral Pion Production $\gamma + p \to \pi^0 + p$

Differential cross sections for π^0-photoproduction from hydrogen have been studied in a series of experiments. As far as the obtained data are concerned, the situation was quite unsatisfactory until recently. This fact was pointed out by *R. L. Walker* at the Daresbury Conference 1969 with the following remark: "...π^0 production from hydrogen, (is) a reaction for which there is a well established tradition of disagreement between different experiments." Since last year, however, a large amount of new data has been reported. A summary of the recent cross section measurements together with some earlier data for energies above 1.5 GeV (which are not listed in Ref. [1]) and all existing data on the asymmetry ratio and on recoil proton polarisation are shown in Fig. 7.

R. L. Walker and *E. B. Wolverton* [12] have taken angular distributions between 574 and 1211 MeV at an average distance of 25 MeV. The angular range at low energies extends from 84° to 170° and from 60° to 170° at energies above 830 MeV. Another set of angular distributions has been measured at Orsay [14] in steps of 50 MeV between 0.7 and 1.15 GeV for angles from 115° to 180°. Besides good agreement *G. Sauvage* [14] noticed deviations at high energies and large angles between the Orsay values and the preliminary data from Caltech. The final results, however, agree very well as can be seen in Fig. 8.

Around 1 GeV a peak at 140° shows again the contribution of the third resonance $F_{15}(1688)$. Likewise good agreement exists with the new data from Daresbury [13] which have been measured at forward angles between 30° and 90° at energies above 0.82 GeV. Excitation curves at 60° and at 90° have been published earlier from CEA [5]. An angular distribution for forward angles has been measured by a Bonn group at DESY [15].

180° cross sections have been investigated at DESY [16] and at Cornell [17]. The results are shown in Fig. 10 together with the data from Orsay [14]. Above 0.8 GeV all three experiments show the same resonance structure but there is a considerable difference in normali-

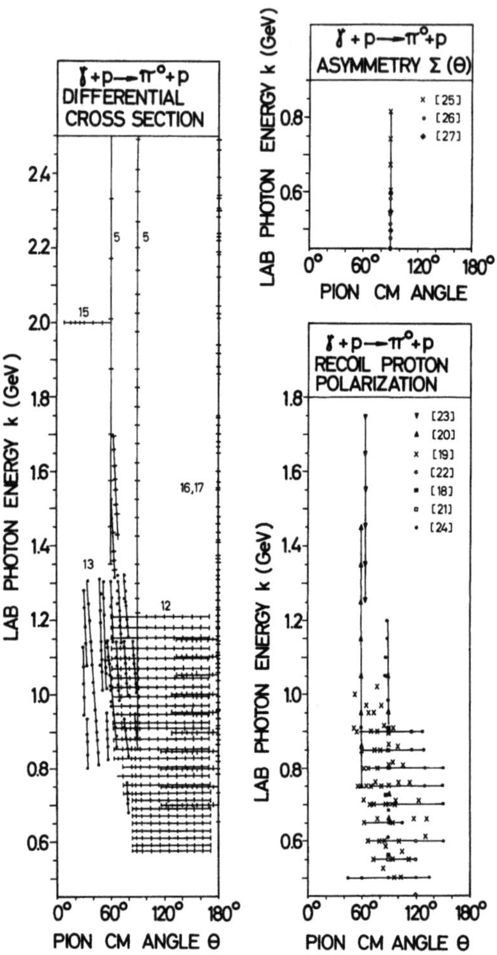

Fig. 7. Summary of experimental data on $\gamma + p \rightarrow \pi^0 + p$. The energies and angles at which measurements of differential cross section, recoil proton polarization and asymmetry have been made are denoted by a point in the θ-k-plane. Angular distributions are shown by horizontal lines, excitation curves by vertical lines. The numbers give the experiments according to the list of references. Only the most recent cross section data which are not listed in Ref. [1] are shown. The angular distributions between 575 and 1210 MeV are from Refs. [12, 14]

sation between the data from DESY and Orsay on one hand and the Cornell values on the other. The marked bump around 1.4 GeV indicates again a strong contribution of the $F_{37}(1920)$.

Polarization of the recoil proton has been studied during the last decade at Frascati [18], Stanford [19], Caltech [20], Cornell [21], Tokyo [22]

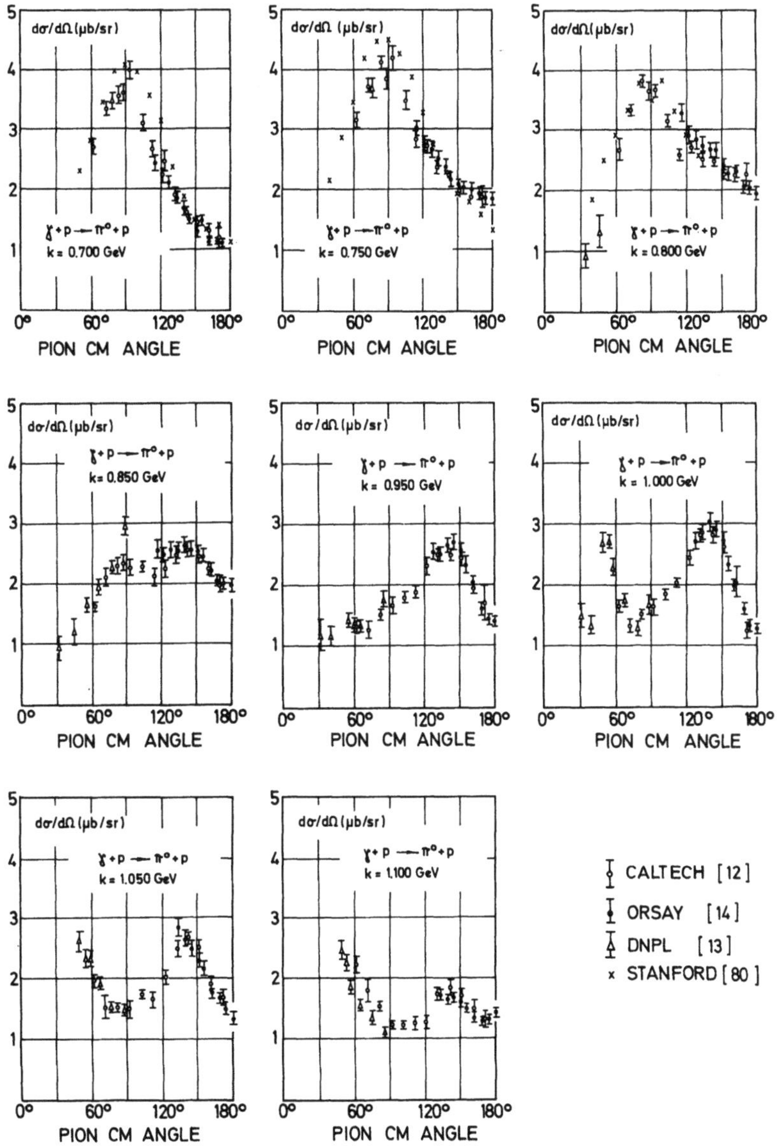

Fig. 8. Angular distributions for $\gamma + p \to \pi^0 + p$ (from Ref. [60])

and Bonn [24]. In Fig. 9 angular distributions of the polarization at energies between 0.5 and 0.9 GeV are plotted using all existing data. The error bars and the rather large fluctuations clearly demonstrate the difficulty of this kind of measurements. The solid lines show the

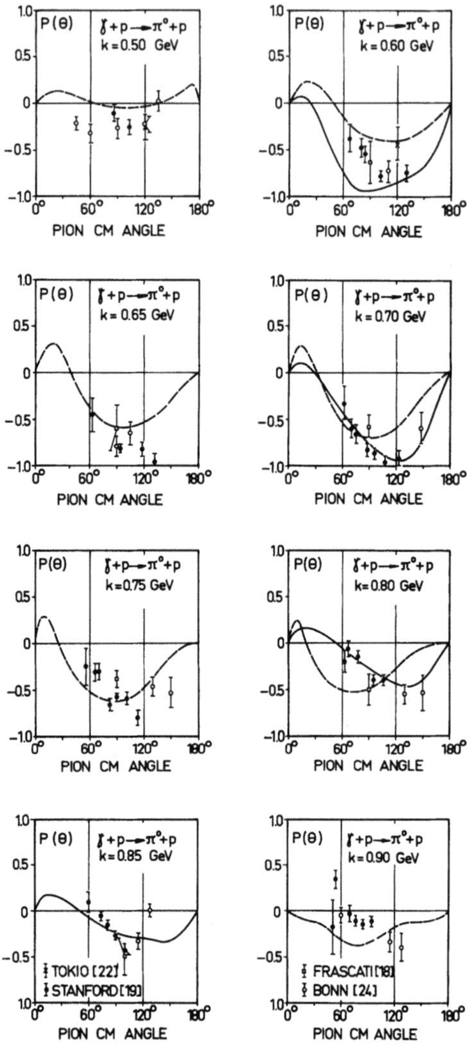

Fig. 9. Angular distributions of recoil proton polarization for $\gamma+p\to\pi^0+p$. The data are from: crosses Ref. [22], squares Ref. [20], open circles Ref. [19], full circles Ref. [24]

fit by *R. L. Walker* [68] the broken line the one by *W. Pfeil* [67] which both are based on the unsatisfactory cross section data available in 1968. A new analysis including the above mentioned new cross section data would probably give better predictions.

At energies above 1.2 GeV new polarization measurements have been reported from CEA [23] at 65°. Fig. 11 shows the data points

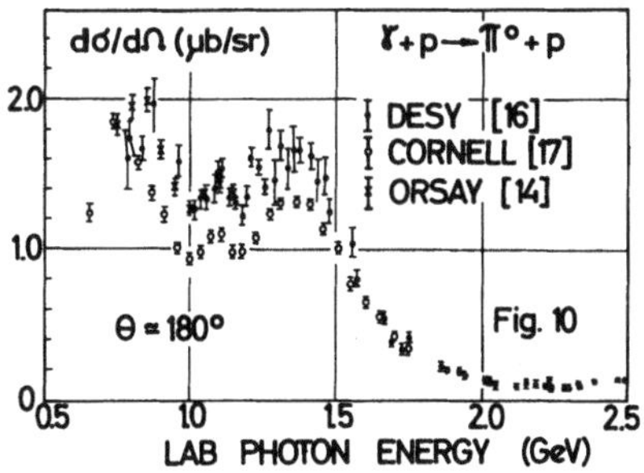

Fig. 10. Energy dependence of 180° differential cross section for $\gamma + p \to \pi^0 + p$

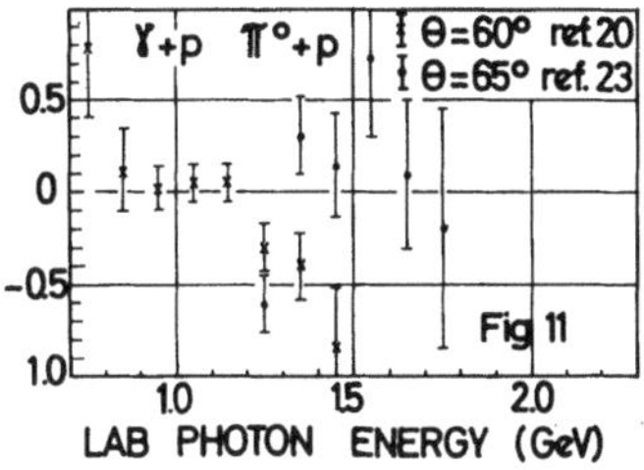

Fig. 11. Energy dependence of recoil proton polarization for $\gamma + p \to \pi^0 + p$ at 60° (Ref. [20], crosses) and 65° (Ref. [23], full circles)

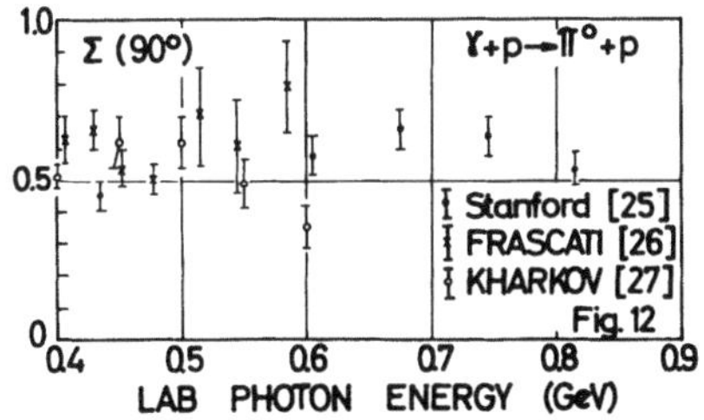

Fig. 12. Asymmetry parameter for linearly polarized photons for $\gamma + p \to \pi^0 + p$ at 90°

together with previous results from Caltech [20] for lower photon energies at 60°. According to the authors the deviations around 1.4 GeV are caused by the fact that the angular dependence of the polarization is very steep and shows a zero crossing at 60°.

π^0 measurements with linearly polarized photons at 90° up to 0.8 GeV have been reported earlier from Stanford [25]. Fig. 12 shows the results together with new asymmetry data from Frascati [26] and Kharkov [27].

3.3. Negative Pion Production $\gamma + n \rightarrow \pi^- + p$

Negative pion production from neutrons above 0.5 GeV has first been studied at Caltech [1, 84] by measuring the π^-/π^+ ratio from deuterium at six different photon energies at several angles between 0.5 and 1.0 GeV. The cross sections have then been derived from the measured π^-/π^+ ratios and the known π^+ photoproduction data.

A summary of the more recent investigations is shown in Fig. 13. *P. E. Scheffler* and *P. L. Walden* [28] at Caltech have measured the energy

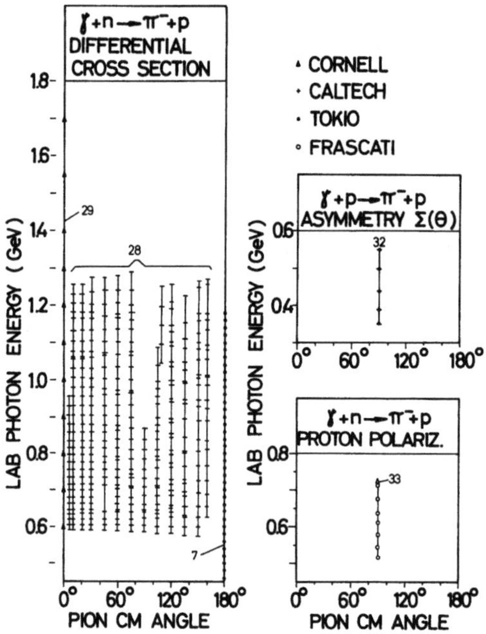

Fig. 13. Summary of experimental data on $\gamma + n \rightarrow \pi^- + p$. The energies and angles at which differential cross section, recoil proton polarization and asymmetry measurements have been made are denoted by a point in the k-θ-plane. Angular distributions are shown by horizontal lines, excitation curves by vertical lines. The numbers give the experiments according to the list of references

dependence of the differential cross section at 13 angles between 6° and 160°. Fig. 14 shows a sample of angular distributions which were obtained by interpolation. The solid lines are Moravczik fits to the data (from Ref. [28]). Similar to π^+ production the cross section exhibits a sharp peak in forward direction, again caused by OPE contributions. The broad peak at large angles, however, indicates a different behaviour in backward production. According to *Scheffler* and *Walden* the total cross section calculated from their data (Fig. 1) shows a peak around 700 MeV

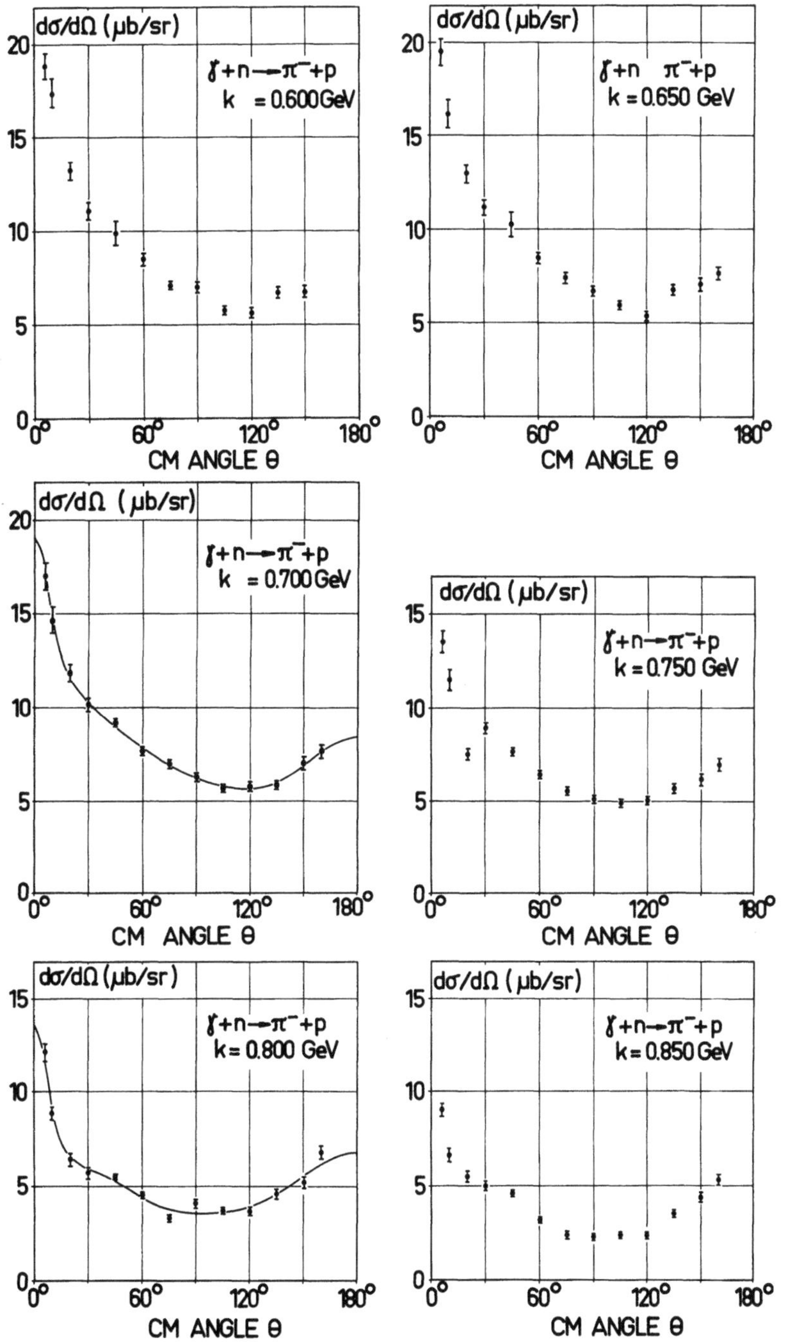

Fig. 14. Angular distributions for $\gamma + n \rightarrow \pi^- + p$ showing data and Moravscik fit curves from Ref. [28]

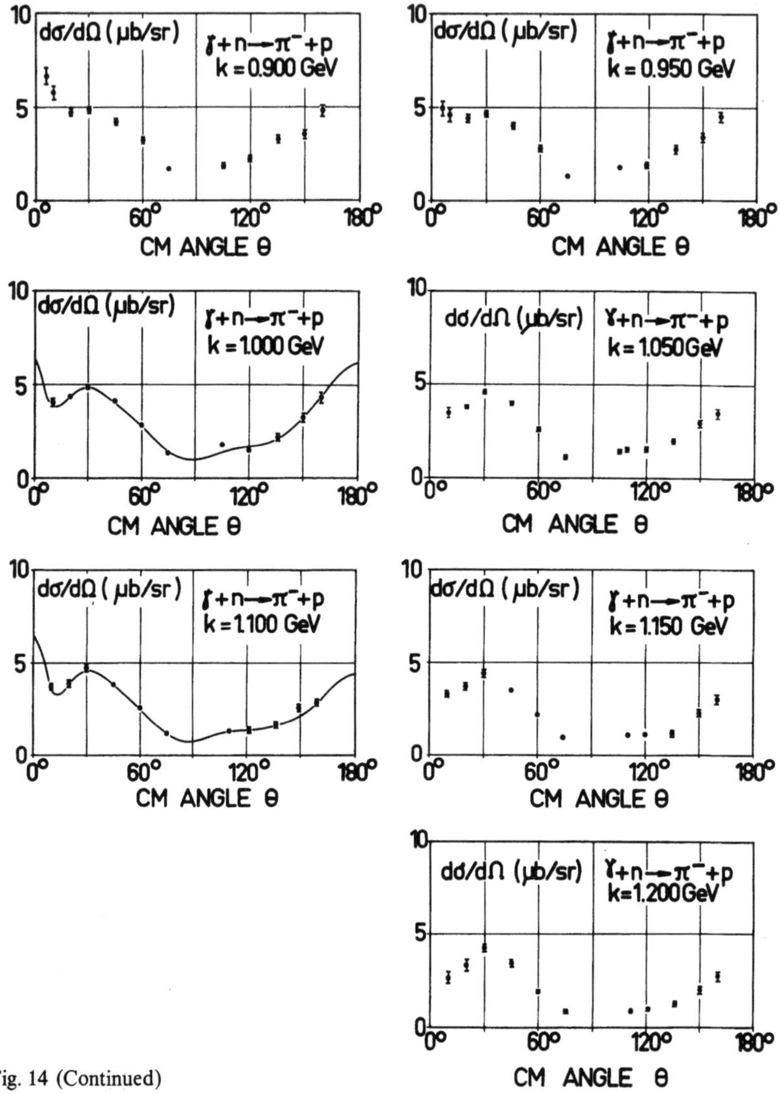

Fig. 14 (Continued)

at the "second resonance" position but no enhancement around 1 GeV. This fact demonstrates that the contribution of the "third resonance" $F_{15}(1688)$ is at least small which is in agreement with recent quark model predictions [85].

The absence of F_{15} contributions is not so clearly demonstrated by the 180° cross sections measured at the Tokyo synchrotron [7] which are shown in Fig. 15.

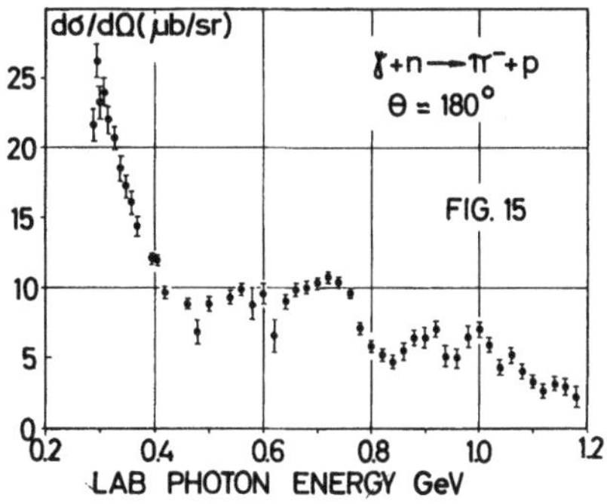

Fig. 15. Energy dependence of differential cross section at 180° for $\gamma+n\rightarrow\pi^-+p$. Data are from Ref. [7]

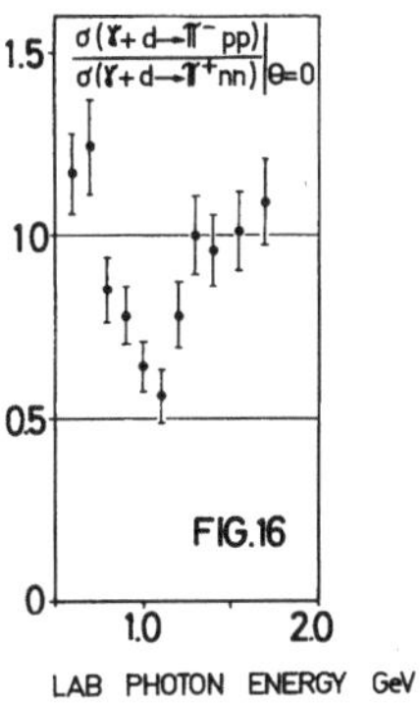

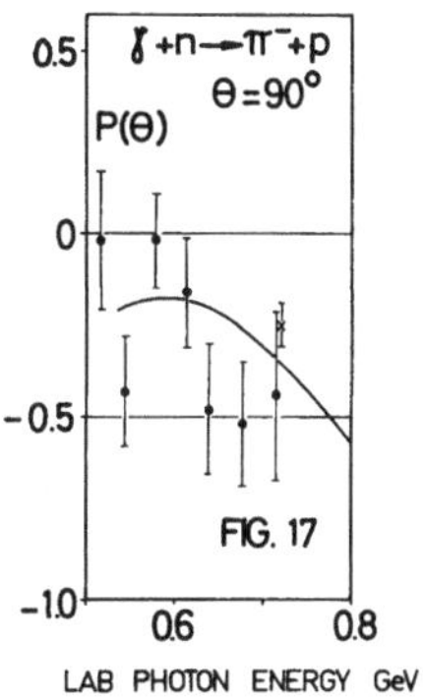

Fig. 16. Ratio of π^+/π^- photoproduction off deuterium versus photon energy showing data from Ref. [29]

Fig. 17. Energy dependence of recoil proton polarization for $\gamma+n\rightarrow\pi^-+p$ at 90°. Data are from Refs. [32] and [33]. The curve is from Ref. [71]

π^-/π^+ ratios at 0° have been measured at Cornell [29] at photon energies between 0.6 and 1.7 GeV. The ratio as plotted in Fig. 16 versus photon energy shows a broad dip around 1.1 GeV.

Preliminary data for total and differential cross sections measured with deuterium bubble chambers have been reported from Frascati [30] for energies from 0.2 to 1.0 GeV and from DESY [31] for 0.2–2.0 GeV.

The polarization of the recoil proton in $\gamma+n\rightarrow\pi^-+p$ at 90° in the incident photon energy range of 0.5 to 0.7 GeV has been measured at Frascati [32]. In Fig. 17 the values are compared with an earlier result and a prediction of *A. Proia* and *F. Sebastiani* [71]. Their analysis is based on the above mentioned DESY deuterium bubble chamber results and the earlier measured polarization at 715 MeV from Cornell [33].

4. Eta Meson Photoproduction $\gamma + p \to \eta + p$

Angular distributions of the differential cross section for $\gamma + p \to \eta + p$ between threshold and 0.9 GeV have been studied at Frascati [38] and at Orsay [41]. The angular range and the respective photon energies are illustrated in Fig. 18 which gives a summary of all existing η-produc-

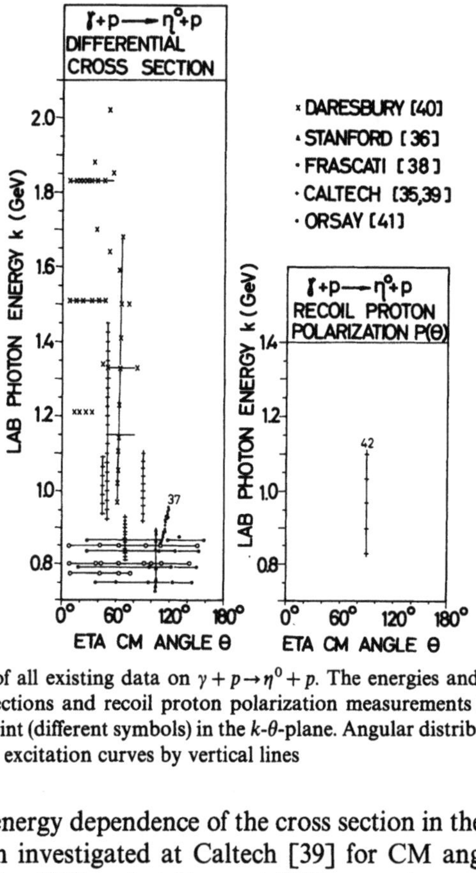

Fig. 18. Summary of all existing data on $\gamma + p \to \eta^0 + p$. The energies and angles at which differential cross sections and recoil proton polarization measurements have been made are denoted by a point (different symbols) in the k-θ-plane. Angular distributions are shown by horizontal lines, excitation curves by vertical lines

tion data. The energy dependence of the cross section in the same energy region has been investigated at Caltech [39] for CM angles of 70°, at Stanford [36] for 105° and at Frascati [37] at angles around 110°. At energies above 0.9 GeV cross section data are reported from Caltech [35, 39] for 45°, 50° and 90°. At Daresbury [40] the energy dependence has been measured around 65° and in addition angular distributions at forward angles for 1.21, 1.51 and 1.83 GeV.

In Fig. 19 all existing data are collected in angular bins of 20° and plotted versus photon energy for CM angles of 50°, 70°, 90° and 110°. The curves are calculated by *R. P. Bajpai* and *A. Donnachie* [75] using a generalized interference model. The characteristic feature of η-produc-

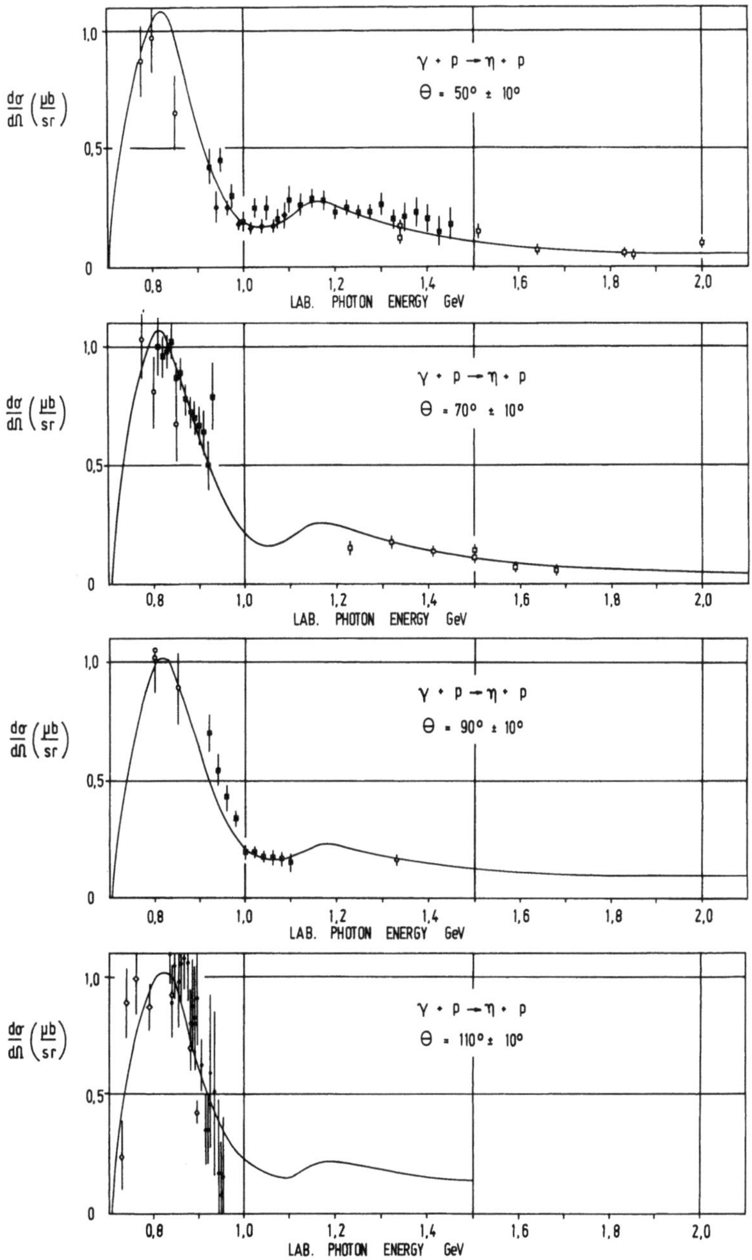

Fig. 19. Energy dependence of differential cross section for $\gamma + p \rightarrow \eta^0 + p$ at 50°, 70°, 90° and 110°. Data are: full squares Ref. [39], full diamonds Ref. [36], open diamonds Ref. [35], full circles Ref. [38] and open squares Ref. [40]. The solid line shows interference model calculations by *R. P. Baijpai* and *A. Donnachie* (from Ref. [75])

tion is a steep rise of the cross section above threshold to a strong peak around 0.82 GeV which is attributed to the $S_{11}(1510)$ resonance. According to recent measurements at higher energies there is also evidence for a second flat maximum near 1.2 GeV which might be due to the excitation of the $P_{11}(1750)$ resonance.

Fig. 20 shows angular distributions for $\gamma + p \rightarrow \eta + p$ for photon energies between 0.75 and 1.83 GeV. At low energies the data show a flat behaviour as we expect from *s*-wave production. The solid lines up to 0.85 GeV are isobar modell fits by *H. G. Beisenherz* [74]. At photon energies above 1.22 GeV some structure shows up in the angular dependence which gives evidence for formation of higher isobars. The curves in this part are likewise obtained from interference model calculations [75].

Measurements of the recoil proton polarization in $\gamma + p \rightarrow \eta + p$ have been done recently at Caltech [42] at 90° for energies between 0.8 and 1.1 GeV which show a sizeable positive polarization in the entire region.

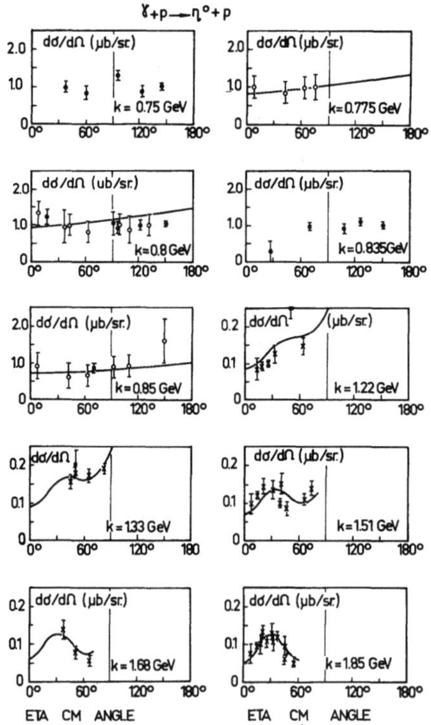

Fig. 20. Angular distributions for $\gamma + p \rightarrow \eta^0 + p$ showing the following data: full circles Ref. [41], open circles Ref. [37, 38], full squares Ref. [35, 39], crosses Ref. [40]. Fit curves from 775 to 850 MeV are from *H. G. Beisenherz* [74] and at higher energies from Ref. [40]

5. Positive Kaon Photoproduction

A large amount of new data have been obtained during the last two years on K^+ production in the $\gamma + p \rightarrow K^+ + \Lambda$ channel as well as for the reaction $\gamma + p \rightarrow K^+ + \Sigma^0$. Fig. 21 gives a summary of the new results together with earlier measurements of the differential cross section and the polarization of the recoil lambda.

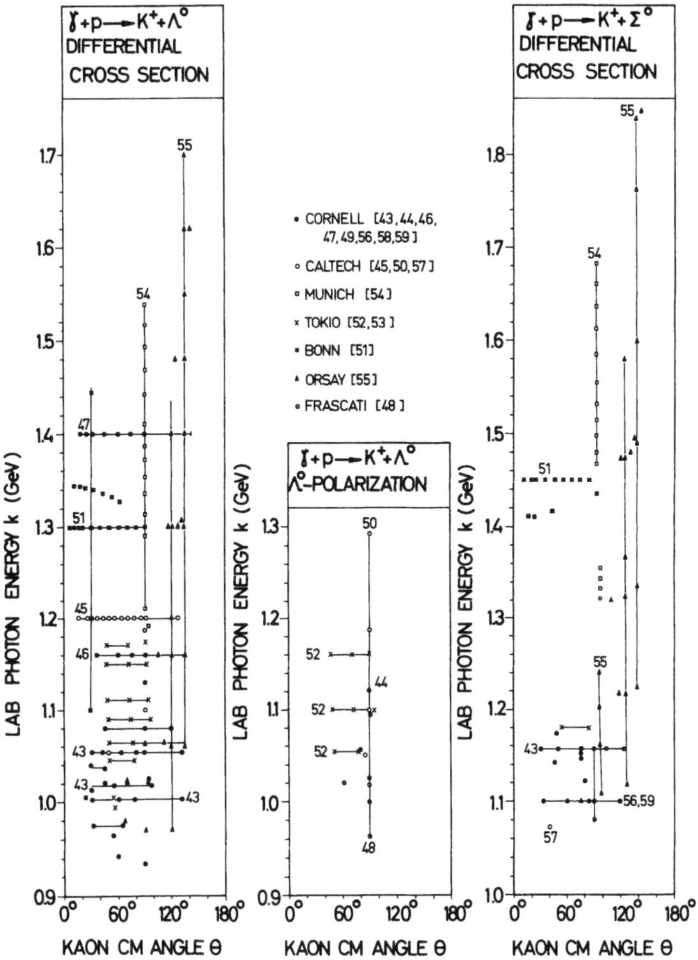

Fig. 21. Summary of all existing data on $\gamma + p \rightarrow K^+ + \Lambda^0$ and $\gamma + p \rightarrow K^+ + \Sigma^0$. The energy and angles at which differential cross sections and polarizations have been measured are denoted by different symbols in the k-θ-plane. Angular distributions are shown by horizontal lines, excitation curves by vertical lines. The numbers give the different experiments according to the list of references

5.1. $\gamma + p \to K^+ + \Lambda$

In a series of measurements [43, 46, 47] at the Cornell synchrotron the angular dependence of the cross section has been measured at photon energies of 1.003, 1.018, 1.08, 1.16 and 1.4 GeV and in addition a few single values between threshold and 1.1 GeV (the data are compiled in Ref. [49]).

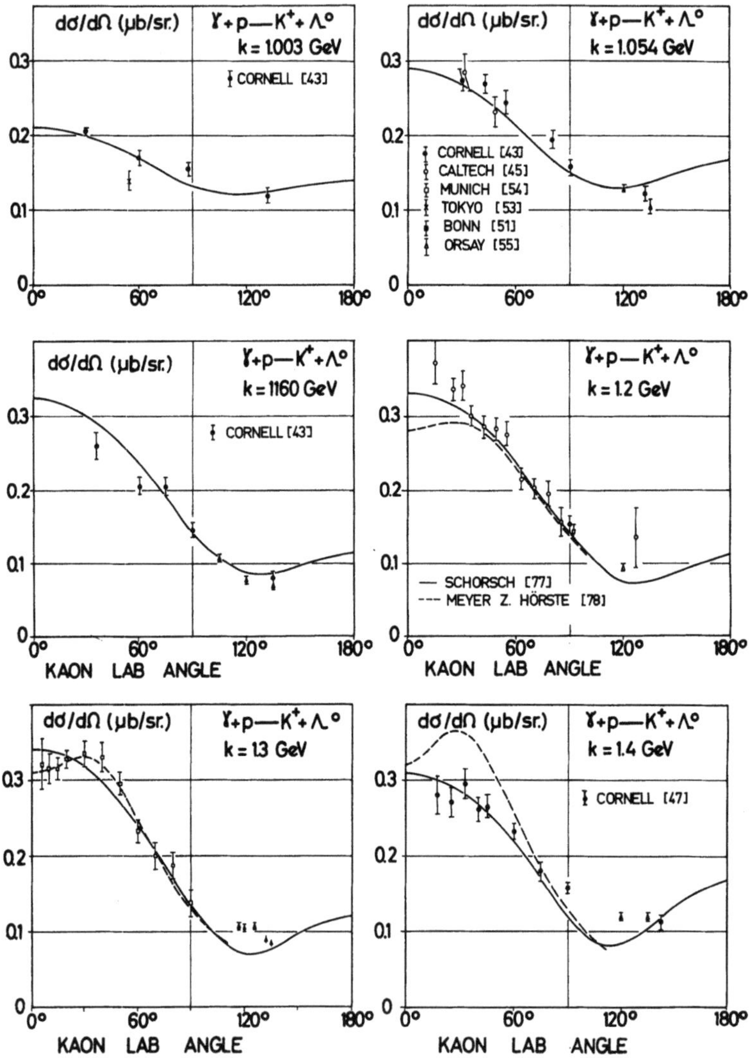

Fig. 22. Angular distributions for $\gamma + p \to K^+ + \Lambda^0$ at six photon energies. The solid curve is the result of an isobar model analysis [77]. The dashed curve shows a Regge model fit [78]

An angular distribution at 1.2 GeV has been published from Caltech [45]. The differential cross section between 46° and 90° at photon energies from 1.0 to 1.17 GeV was studied at Tokyo [53]. An excitation curve at 30°, an angular distribution at 1.3 GeV and some more forward points around 1.32 GeV have been measured at Bonn [51]. The energy dependence around 90° from 1.19 to 1.54 GeV was investigated at DESY [54]. Backward measurements up to kaon angles of 138° in the energy range from 0.96 to 1.7 GeV have been published from Orsay [55] recently.

Fig. 22 shows a sample of angular distributions. The solid lines are fit curves from a multipole analysis by *W. Schorsch, J. Tiedge* and *W. Weilnböck* [77] which is similar to the earlier analysis of *S. Orito* [76]. The broken line gives Regge model calculations from *V. Meyer zu Hörste* and *W. Pfeil* [78] for forward angles.

In Fig. 23 the measured cross sections are plotted versus photon energy for 30°, 90°, 120° and 138°. The recent large angle data from Orsay show a bump structure of $(K^+ + \Lambda)$ production which is also indicated by the 30° excitation curve from Bonn.

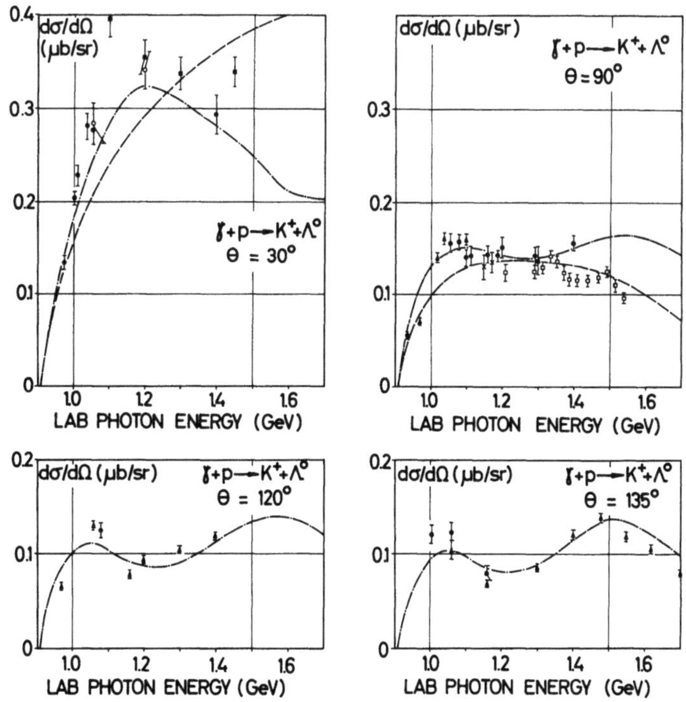

Fig. 23. Energy dependence of differential cross section for $\gamma + p \rightarrow K^+ + \Lambda^0$ at 30°, 90°, 120° and 135°. The curves show fits using isobar model [79] and Regge model [78]

A little while ago there was no clear indication for a resonance behaviour in this reaction. Only the analysis of the recoil lambda polarization required a $P_{11}(1700)$ resonance contribution [76]. According to *F. M. Renard* and *Y. Renard* [79] the two bumps appearing at large angles for $E_\gamma = 1$ GeV and $E_\gamma = 1.5$ GeV are due to interference between Born terms and $D_{13}(1520)$, $F_{15}(1688)$ and $P_{11}(1785)$ for the first bump and $P_{13}(1855)$ and $D_{13}(2030)$ for the second one. The dashed lines give the result of their analysis.

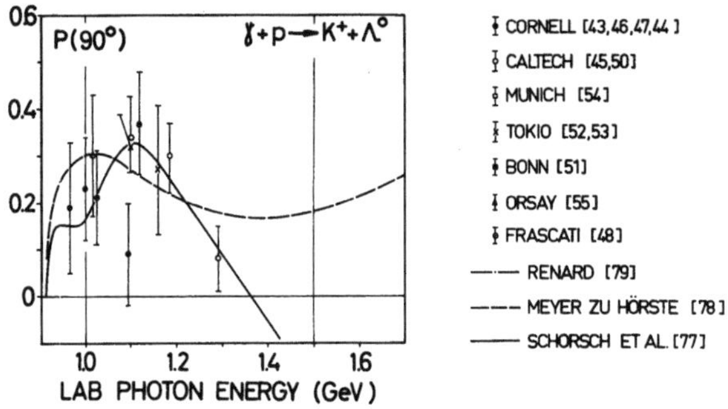

Fig. 24. Polarization of recoil lambda for $\gamma + p \to K^+ + \Lambda^0$ at 90°. Solid line shows isobar model fit [77], broken line Regge model analysis [78]

The polarization of the lambda particle in $\gamma + p \to K^+ + \Lambda$ at 90° and at energies between threshold and 1.3 GeV has been measured at Cornell [44], Caltech [50] and Frascati [48]. Investigations of the angular dependence of $P(\theta)$ from 46° to 90° at energies of 1.054, 1.1 and 1.16 GeV have been carried out at Tokyo [52]. Fig. 24 shows the energy dependence of the polarization at 90° together with fit curves from Refs. [77, 78].

5.2. $\gamma + p \to K^+ + \Sigma^0$

Measurements of the angular dependence of the cross section for $\gamma + p \to K^+ + \Sigma^0$ have been done in the early sixties at Cornell [43, 56, 59] for photon energies of 1.1 and 1.16 GeV. Single values below 1.2 GeV photon energy are also reported from Caltech [57] and Tokyo [53]. An angular distribution from 10° to 85° at 1.45 GeV was studied at Bonn [51]. An excitation curve of the cross section at 90° between $E_\gamma = 1.32$ GeV and $E_\gamma = 1.68$ GeV has been measured by a Munich group at DESY [54]. The energy dependence of the cross section around 95°, 124° and 138° has recently been published from Orsay [55].

The three existing angular distributions are shown in Fig. 25. The curves give the isobar model fits of *W. Schorsch*, *J. Tiedge* and *W. Weilnböck*

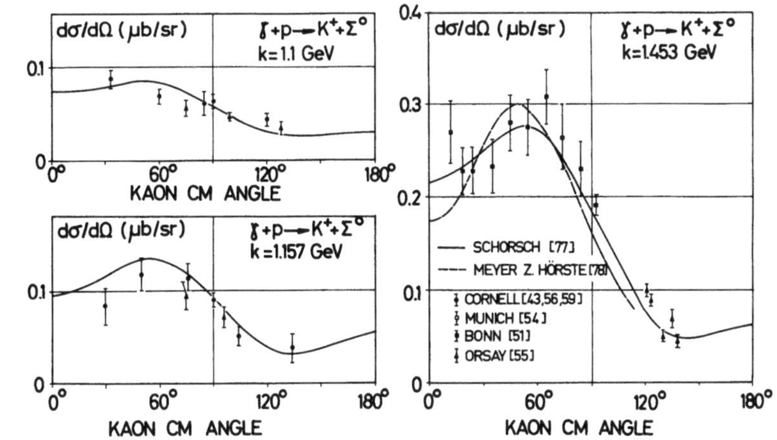

Fig. 25. Angular distributions for $\gamma + p \to K^+ + \Sigma^0$. Solid line shows isobar model analysis [77], the dashed line Regge model fit [78]

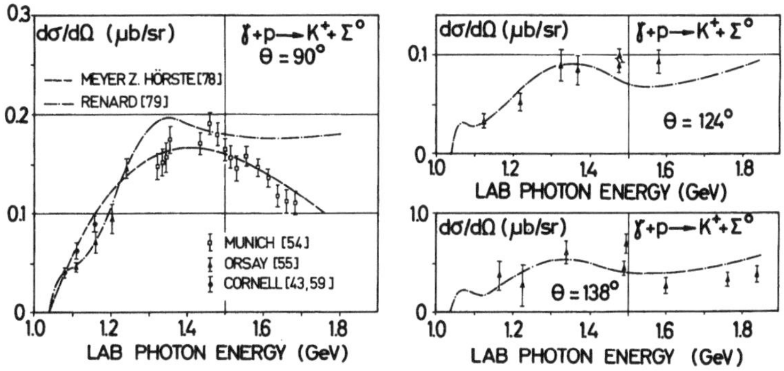

Fig. 26. Energy dependence of differential cross section for $\gamma + p \to K^+ + \Sigma^0$ at 90°, 124° and 138°. The curves are from an isobar model analysis [79] and from Regge model fit [78]

[77] (solid lines) and the Regge model calculations of *V. Meyer zu Hörste* and *W. Pfeil* [78] (broken lines).

The energy dependence of the differential cross section for 90°, 124° and 138° is plotted in Fig. 26 together with fit curves of *V. Meyer zu Hörste* and *W. Pfeil* [78] and *F. M. Renard* and *Y. Renard* [79]. At 90° there is a clear resonance behaviour which is less evident at larger angles. The maximum of the cross section is centered around 1.45 GeV which corresponds to a total CM energy of about 1.9 GeV. In the $(K^+ + \Sigma^0)$ channel resonances with isospin $I = 1/2$ and $I = 3/2$ can contribute in contrast to the $(K^+ + \Lambda)$ channel. Therefore we have several candidates e.g. $P_{13}(1860)$, $F_{35}(1890)$, $P_{31}(1910)$ and $F_{37}(1950)$. In general the F_{37} (1950) is favored, but there are not enough data for a reliable assignment.

6. Summary

The recent experimental achievements in intermediate energy photoproduction of pseudoscalar mesons may be summarized as follows:

a) Positive pion production has been predominantly measured at large angles. At 180° there are large deviations between four different experiments which makes the situation quite unclear. For the same reaction we have obtained new information on recoil neutron polarisation.

b) The knowledge of the differential cross sections in π^0-production has been very much improved by three experiments. We have obtained also new information on recoil proton polarization over a wide range of angle and energy. New data for the asymmetry ratio $\Sigma(\theta)$ at 90° have been published too.

c) A new set of angular distributions for π^- production from deuterium and also the measurement of an excitation curve at 180° has very much improved the data for this reaction. Also measurements of the polarization have been made.

d) New values for the cross section and the polarization in η-production have given new information about the resonances involved in this reaction.

e) Several measurements of K^+-production show considerable structure in the energy dependence of the cross section as well in the $(K^+ + \Lambda)$ channel as in the $(K^+ + \Sigma^0)$ channel. This fact has been suggested for a long time but was not clearly seen in experimental data.

The data from the previous measurements together with these new results give the basis for our present knowledge of the resonance behaviour of meson photoproduction. The main features may be pointed out in a few words.

The excitation of the dominant resonances $P_{33}(1236)$, $D_{13}(1520)$ and $F_{15}(1688)$ in pion photoproduction is well established. According to recent measurements however the $F_{15}(1688)$ contributes weakly to $\gamma + n \rightarrow \pi^- + p$. The presence of all those resonances which give small contributions can only be noticed by carring out a multipole analysis. In all such calculations so far the $P_{11}(1470)$ and also the $D_{15}(1670)$ had to be included in order to obtain a good fit to π production data.

Most uncertain in this reaction are the contributions of the s-wave resonances. The $S_{11}(1535)$ seems fairly well established. In the calculations which also covered the region of the third resonance satisfactory fits could be obtained with and without the $S_{31}(1650)$ and $S_{11}(1700)$ and also by including only one of the two. The existence of the $D_{33}(1670)$ in π-N-scattering has been controversial for a long time, therefore no attempt has been made for their identification in photoproduction.

At energies above the third resonance no systematic analysis of pion production could be made for lack of data. Strong evidence for a $F_{37}(1950)$ excitation gives the bump structure of the cross sections especially at backward directions.

All analyses of eta production show a strong contribution of the $S_{11}(1535)$ which again appears as a marked bump in the cross section. Different opinions exist about the resonances at higher energies. Fits to the data have been obtained by including the $D_{15}(1670)$ or the D_{13} (1520) and the $F_{15}(1688)$ or the $P_{11}(1785)$.

In $\gamma + p \to K^+ + \Lambda$ production the resonance contributions are found to be weak. All calculations require a $P_{11}(1780)$. In some analyses an additional small contribution of the $S_{11}(1700)$ is needed. In one case also a $D_{13}(1750)$ amplitude has been noticed. In most cases no contributions are found from $D_{15}(1670)$ and $F_{15}(1688)$ which both are important in π production. The only exception is the most recent analysis which already included the new backward data and which showed contributions from $S_{11}(1535)$, $D_{13}(1520)$, $F_{15}(1688)$, $P_{13}(1860)$ and $D_{13}(2040)$.

In contrary to $K\Lambda$ reaction there are considerable resonance contributions in $\gamma + p \to K^+ + \Sigma^0$. In a recent analysis fits to the cross sections could be obtained with contributions of $S_{11}(1700)$, $P_{31}(1910)$ and F_{35} (1890) or $F_{37}(1950)$. For this reaction however threre are less experimental data than for $\gamma + p \to K^+ + \Lambda$ therefore the results are inevitably less significant.

Despite the large number of measurements carried out so far our understanding of intermediate energy photoproduction is still fragmentary. To improve this situation, more accurate data are needed especially at higher energies.

References

1. *Beale, J. T., Eklund, S. D., Walker, R. L.*: California Institute of Technology, Synchrotron Laboratory Report CTSL–42 (1968).
2. *Betourne, C., Bizot, J. C., Perez-Y-Jorba, J., Treille, D., Schmidt, W.*: Phys. Rev. **172**, 1343 (1968).
3. *Buschhorn, G., Caroll, J., Eandi, R. D., Heide, P., Hübner, R., Kern, W., Kötz, U., Schmüser, P., Skronn, H. J*: Phys. Rev. Letters **17**, 1027 (1966), and **18**, 571 (1967).
4. *Alvarez, R. A., Cooperstein, G., Kalata, G., Lanza, R. C., Luckey, D.*: Phys. Rev. D**1**, 1946 (1970)
5. *Alvarez, R. A., Bar-Yam, Z., Kern, W., Luckey, D., Osborne, L. S., Tazzari, S.*: Phys. Rev. Letters **12**, 707 (1964).
6. *Ekstrand, K., Browman, A., Hand, L., Nordberg, E.*: Cornell University Preprint CLNS–124 (1970).
7. *Fujii, T., Okuno, H., Orito, S., Sasaki, H., Nozaki, T., Tasaki, F., Takikawa, K., Amado, K., Endo, I., Yoshida, K., Higuchi, M., Sato, M., Sumi, Y.*: University of Tokio INS Report 149 (1970).

8. *Fischer, G., Fischer, H., Leneke, M., Wessels, H.:* Preprint Bonn University PI 1–100 (1970).
9. *D'Almagne, B.:* Orsay Thesis L.A.L. 1239 (1970). – *Bouquet, B., Croissiaux, M., d'Almagne, B., Erickson, E., Eschruth, P., Kitching, P., Morand, R., Pahin, J.P., Woulds, J.:* To be published.
10. *Hahn, U., Wallraff, W.:* Bonn University Report PI 1–87 (1970).
11. *Liu, F.F., Vitale, S.:* Phys. Rev. **144**, 1093 (1966).
12. *Wolverton, F.B.:* Ph. D. Thesis, California Institute of Technology (1968). Preliminary data. – *Walker, R.L., Wolverton, F.B.:* To be published.
13. *Booth, P.S.L., Butler, M.F., Carrol, L.J., Holt, J.R., Jackson, J.N., Range, W.H., Williams, E.G.H., Wormald, J.R., Tait, N.R.S.:* Private communication.
14. *Sauvage, G.:* Orsay Thesis (1969) L.A.L. 1207. – *Delcourt, B., Lefrancois, J., Parrour, G., Perez-Y-Jorba, J.P., Sauvage, G.:* Phys. Letters **29** B, 70 (1969).
15. *Braunschweig, M., Braunschweig, W., Husmann, D., Lübelsmeyer, K., Schmitz, D.:* Nucl. Phys. B**20**, 191 (1970).
16. *Buschhorn, G., Heide, P., Kötz, U., Lewis, R.A., Schmüser, P., Skronn, H.J.:* Phys. Rev. Letters **20**, 230 (1968).
17. *Cassiday, G.L., Fischer, H., Ito, A., Loh, E.C., Rutherfoord, J.:* Phys. Rev. Letters **21**, 933 (1968).
18. *Querzoli, R., Salvini, G., Silvermann, A.:* Nuovo Cimento **19**, 53 (1961).
19. *Lundquist, D.E., Anderson, R.L., Allaby, J.V., Ritson, D.M.:* Phys. Rev. **168**, 1527 (1968). – *Maloy, J.O., Peterson, V.Z., Salandin, G.A., Waldner, F., Manfredini, A., Kendall, H.:* Phys. Rev. **139**, B 733 (1965).
20. *Cheng, S.U., Prescott, C.Y., Heusch, C.A.:* Contribution to Daresbury Conference 1969, Abstract No. 60. – *Bloom, E.D., Heusch, C.A., Prescott, C.Y., Rochster, L.S.:* Phys. Rev. Letters **19**, 671 (1967).
21. *Stein, P.C.:* Phys. Rev. Letters **2**, 473 (1959).
22. *Hayakawa, S., Horikawa, N., Kajikawa, R., Kikuchi, K., Kobayakawa, H., Masaike, A., Mori, K., Obayashi, H., Ukai, K., Yosdida, K.:* J. Phys. Soc. Jap. **25**, 307 (1968).
23. *Deutsch, M., Patel, D., Tsipis, C., Castro, M., Milburn, R., Richards, W., Rutherfoord, J., Sinclair, C., Stearns, B., Tanaka, N.:* Contribution to Daresbury Conference 1969, Abstract No. 13.
24. *Blüm, P., Brinckmann, P., Brockmann, R., Lütter, P., Mohr, W., Feller, P., Minatti, N., Minatti, N., Piel, H., Niehaus, W., Trines, D.:* Bonn University Preprint PI 1–105 (1970) and private communication.
25. *Mozley, R.F.:* Proceedings of the International Conference on Electromagnetic Interactions at Low and Intermediate Energies, Dubna 1967.
26. *Bologna, G., Fabbri, F.L., Spillantini, P., Valente, V.:* Frascati Report LNF–70/39 (1970).
27. *Antufyev, Y.P., Agranovich, V.L., Ganenko, V.B., Gorbenko, V.G., Zhebrovsky, Y.V., Kolesnikov, L.Y., Kuzmenko, V.S., Miroshnichenko, I.I., Sorokin, P.V., Rubashkin, A.L., Chechetenko, V.F.:* Physico-Technical Institute Kharkov Preprint, Contribution to the XV. Conference on High Energy Physics, Kiev 1970.
28. *Scheffler. P.E., Walden, P.L.:* Phys. Rev. Letters **24**, 952 (1970), and private communication.
29. *Ito, A., Loe, R., Loh, E., Ramanauskas, A., Ritchie, D., Schmidt, W.:* Cornell University Report CLNS–92 (1970).
30. *Lodi-Rizzini, E., Mantovani, G.C., Piazzoli, A., Fiore, L., Gialanella, G., Rossi, V., Piazza, A., Susinno, G., Carbonaro, F., Napolitano, M., Rinzivillo, R.:* Lett. Nuovo Cimento **3**, 697 (1970).
31. *Hilpert, H.G., Lauscher, P., Matziolis, M., Schnackers, H., Weber, H., Meyer, A., Pose, A., Böckmann, K., Idschock, U., Müller, K., Paul, E., Propach, E., Butenschön, H., Kübeck, H., Lüke, D., Seebeck, H., Spitzer, H., Stormm, F., Brandt, S., Braun, O., Steffen, P., Stiewe, J., Schlamp, P., Weigl, J., Wilkinson, K.:* Nucl. Phys. B **8**, 535 (1968).

32. *Beneventano, M., d'Angelo, S., de Notaristefani, F., Monacelli, P., Paoluzi, L., Sebastiani, F., Severi, M., Stelle, B.*: Lett. Nuovo Cimento **3**, 840 (1970).
33. *Kenemuth, J. R., Stein, P. C.*: Phys. Rev. **129**, 2259 (1963).
34. *Nishikawa, T., Hiramatsu, S., Kimura, Y., Kobayashi, M., Kondo, K., Okumura, S., Suzuki, T., Takikawa, K., Tsuru, T., Yoshida, H.*: Phys. Rev. Letters **21**, 1288 (1968).
35. *Heusch, C. A., Prescott, C. Y., Bloom, E. D., Rochester, L. S.*: Phys. Rev. Letters **17**, 573 (1966).
36. *Prepost, R., Lundquist, D., Quinn, D.*: Phys. Rev. Letters **18**, 82 (1967).
37. *Bacci, C., Penso, G., Salvini, G., Mencuccino, C., Silvestrini, V.*: Nuovo Cimento **45** A, 983 (1966), and Phys. Rev. Letters **16**, 157 (1966).
38. *Bacci, C., Baldini-Celio, R., Mencuccino, C., Reale, A., Spinetti, M., Zallo, A.*: Phys. Rev. Letters **20**, 571 (1968).
39. *Bloom, E. D., Heusch, C. A., Prescott, C. Y., Rochester, L. S.*: Phys. Rev. Letters **21**, 1100 (1968).
40. *Booth, B. S. L., Butler, M. F., Carroll, L. J., Holt, J. R., Jackson, J. N., Range, W. H., Williams, E. G. H., Wormald, J. R., Tait, N. R. S.*: Lett. Nuovo Cimento **2**, 66 (1969), and University of Liverpool Preprint Oct. 1970. To be published in Nucl. Phys. B.
41. *Delcourt, B., Lefrancois, J., Perez-Y-Jorba, J. P., Sauvage, G., Menuessier, G.*: Phys. Letters **29** B, 75 (1969).
42. *Heusch, C. A., Prescott, C. Y., Rochester, L. S., Winstein, B. D.*: Contribution to the XV. International Conference on High Energy Physics, Kiev 1970.
43. *Anderson, R. L., Gabathuler, E., Jones, D., McDaniel, B. D., Sadoff, A. J.*: Phys. Rev. Letters **11**, 433 (1963).
44. *Thom, H., Gabathuler, E., Jones, D., McDaniel, B. D., Sadoff, A. J.*: Phys. Rev. Letters **11**, 433 (1963).
45. *Peck, C. W.*: Phys. Rev. **135**, B 830 (1964).
46. *Sadoff, A. J., Anderson, R. L., Gabathuler, E., Jones, D.*: Bull. Am. Phys. Soc. **9**, 34 (1964).
47. *Anderson, R. L., Gabathuler, E., Jones, D., McDaniel, B. D., Mistry, N. B., Mori, S., Sadoff, A. J., Stanton, N. R.*: Proceedings of the International Symposium on Electron and Photon Interactions at High Energies, Hamburg **2**, 203 (1965).
48. *Grilli, M., Mezeti, L., Nigro, M., Schiavuta, E.*: Nuovo Cimento **38**, 1467 (1965).
49. *Thom, H.*: Phys. Rev. **151**, 1322 (1966).
50. *Groom, D. E., Marshall, H. J.*: Phys. Rev. **159**, 1213 (1967).
51. *Bleckmann, A., Herda, S., Opera, U., Schulz, W., Schwille, W. J., Urbahn, H.*: Z. Physik **239**, 1 (1970), and private communication.
52. *Fujii, T., Imanishi, A., Iwata, S., Kusemegi, A., Mishina, M., Miyaji, T., Sasaki, H., Takamatsu, K., Orito, S., Takasaki, F., Ameno, T., Higuchi, M., Homma, S.*: Proc. Int. Symp. on Electron and Photon Interactions Daresbury p. 274 (1969), Phys. Rev. D **2**, 439 (1970).
53. *Iwata, S.*: University of Tokio Preprint INS J 115 (1970).
54. *Göing, H., Schorsch, W., Tietge, J., Weilnböck, W.*: Proc. Int. Symp. on Electron and Photon Interactions Daresbury, p. 302 (1969).
55. *Decamp, D., Dudelzak, B., Eschruth, P., Fourneron, Th.*: Orsay Report L.A.L. 1236 (1970).
56. *McDaniel, B. D., Silvermann, A., Wilson, R. R., Cortellessa, G.*: Phys. Rev. **115**, 1039 (1959).
57. *Brody, H. M., Wetherell, A. M., Walker, R. L.*: Phys. Rev. **119**, 1710 (1960).
58. *Anderson, R. L., Turkot, F., Woodward, W. M.*: Phys. Rev. **123**, 1003 (1961).
59. *Gabathuler, E., Mori, S., Stanton, N.*: Bull. Am. Phys. Soc. **9**, 23 (1964).
60. *Walker, R. L.*: California Institute of Technology CALT-68-219; Proceedings of the International Symposium on Electron and Photon Interaction at High Energies, Daresbury p. 23 (1969).
61. *Noelle, P., Pfeil, W., Schwela, D.*: Bonn University Preprint PI 2–79 (1970).
62. *Berends, F. A., Donnachie, A., Weaver, D.*: Nucl. Phys. B **4**, 1.

63. *Berends, F. A., Donnachie, A.*: University of Glasgow Preprint, Contribution to the XIV. International Conference on High Energy Physics, Vienna 1968.
64. *Schmidt, W., Schwiderski*: Report 3/67–1, 1968 Kernforschungszentrum Karlsruhe.
65. *Yamaki, T.*: Progr. Theor. Phys. **38**, 153 (1967).
66. *Chau, Y. C., Dombey, N., Moorhouse, R. G.*: Phys. Rev. **163**, 1632 (1967).
67. *Pfeil, W., Kim Yong Taik*: University of Bonn Report PI 2–46 and PI 2–51 (1968).
68. *Walker, R. L.*: Phys. Rev. **182**, 1729 (1969).
69. *Moorhouse, R. G., Rankin, W. A.*: Nucl. Phys. B **23**, 181 (1970).
70. *Noelle, P., Pfeil, W.*: Bonn University Preprint PI 2–87 (1971).
71. *Proia, A., Sebastiani, F.*: Lett. Nuovo Cimento **3**, 483 (1970).
72. *Deans, S. R., Holladay, W. G.*: Phys. Rev. **161**, 1466 (1967).
73. *Srinivasan, S. K., Achuthan, P., Sarkar, R. N.*: Nuovo Cimento **59** a, 171 (1969).
74. *Beisenherz, H. G.*: University of Bonn Report PI 2–62 (1969).
75. *Bajpai, R. P., Donnachie, A.*: Nucl. Phys. B **12**, 274 (1969).
76. *Orito, S.*: University of Tokyo Report INS 134 (1969).
77. *Schorsch, W., Tietge, J., Weilnböck, W.*: Nucl. Phys. B **25**, 179 (1971).
78. *Meyer zu Hörste, V., Pfeil, W.*: University of Bonn Report PI 2–85 (1970), to be published.
79. *Renard, F. M., Renard, Y.*: Preprint PM/70/2; University of Montpellier, France.
80. *DeStaebler jr., H., Erickson, E. F., Hearn, A. C., Schaerf, C.*: Phys. Rev. **140**, B 336 (1965).
81. *Hand, L., Schaerf, C.*: Phys. Rev. Letters **6**, 229 (1961). – *Schaerf, C.*: Nuovo Cimento **44**, 504 (1966).
82. *Thiessen, H. A.*: Phys. Rev. **155**, 1488 (1967).
83. *Fischer, G., v. Holtey, G., Knop, G., Stümpfig, J.*: Preprint Bonn University PI 1–101 (1970).
84. *Neugebauer, G., Wales, W., Walker, R. L.*: Phys. Rev. **119**, 1726 (1960).
85. *Copley, L. A., Karl, G., Obryk, E.*: Nucl. Phys. B **13**, 303 (1969).
86. *Spillantini, P., Valente, V., Nigro, M., Oleari, C.*: Nucl. Phys. B **13**, 320 (1969).
87. *Höhler, G.*: Proceedings of the International Symposium on Electron and Photon Interactions at High Energies. Springer Tracts Mod. Phys. **39**, 55 (1965).
88. *Rollnik, H.*: Proceedings of the Heidelberg International Conference on Elementary Particles, p. 400 (1967).
89. Particle Data Group; Phys. Letters **33** B, 9 (Aug. 1970).
90. *Althoff, K., Herr, G. H., Hoffmann, W., Kadansky, V., Opara, U., Schittko, F. J., Schulz, W., Schwille, W. J.*: Private communication.
91. *Joos, P.*: DESY-HERA 10–1 (September 1970).

Dr. *H. Fischer*
Physikalisches Institut der Universität
5300 Bonn, Germany

MIX
Papier aus verantwortungsvollen Quellen
Paper from responsible sources
FSC® C105338

If you have any concerns about our products,
you can contact us on
ProductSafety@springernature.com

In case Publisher is established outside the EU,
the EU authorized representative is:
Springer Nature Customer Service Center GmbH
Europaplatz 3, 69115 Heidelberg, Germany

Printed by Libri Plureos GmbH
in Hamburg, Germany